AS in a week

Thane Gilmour and
Cambridge
Kevin Byrne

Physics

Where to find the information you need

SUCCESS OR YOUR MONEY BACK

Letts' market leading series AS in a Week gives you everything you need for exam success. We're so confident that they're the best revision books you can buy that if you don't make the grade we will give you your money back!

HERE'S HOW IT WORKS

Register the Letts AS in a Week guide you buy by writing to us within 28 days of purchase with the following information:

- Name
- Address
- Postcode
- Subject of AS in a Week book bought

Please include your till receipt

To make a **claim**, compare your results to the grades below. If any of your grades qualify for a refund, make a claim by writing to us within 28 days of getting your results, enclosing a copy of your original exam slip. If you do not register, you won't be able to make a claim after you receive your results.

CLAIM IF...

You are an AS (Advanced Subsidiary) student and do not get grade E or above.
You are a Scottish Higher level student and do not get a grade C or above.
This offer is not open to Scottish students taking SCE Higher Grade, or Intermediate qualifications.

Registration and claim address:
Letts Success or Your Money Back Offer, Letts Educational, 414 Chiswick High Road, London W4 5TF

TERMS AND CONDITIONS

1. Applies to the Letts AS in a Week series only
2. Registration of purchases must be received by Letts Educational within 28 days of the purchase date
3. Registration must be accompanied by a valid till receipt
4. All money back claims must be received by Letts Educational within 28 days of receiving exam results
5. All claims must be accompanied by a letter stating the claim and a copy of the relevant exam results slip
6. Claims will be invalid if they do not match with the original registered subjects
7. Letts Educational reserves the right to seek confirmation of the level of entry of the claimant
8. Responsibility cannot be accepted for lost, delayed or damaged applications, or applications received outside of the stated registration/claim timescales
9. Proof of posting will not be accepted as proof of delivery
10. Offer only available to AS students studying within the UK
11. SUCCESS OR YOUR MONEY BACK is promoted by Letts Educational, 414 Chiswick High Road, London W4 5TF
12. Registration indicates a complete acceptance of these rules
13. Illegible entries will be disqualified
14. In all matters, the decision of Letts Educational will be final and no correspondence will be entered into

Letts Educational
Chiswick Centre
414 Chiswick High Road
London W4 5TF
Tel: 020 8996 3333
Fax: 020 8743 8390
e-mail: mail@lettsed.co.uk
website: www.letts-education.com

Every effort has been made to trace copyright holders and obtain their permission for the use of copyright material. The authors and publishers will gladly receive information enabling them to rectify any error or omission in subsequent editions.

First published 2000
New edition 2004
10 9 8 7 6 5 4 3 2

Text © Thane Gilmour and Glenn Hawkins 2000
Design and illustration © Letts Educational Ltd 2000

British Library Cataloguing in Publication Data
A CIP record for this book is available from the British Library.

ISBN 1 84315 358 0

Cover design by Purple, London

Prepared by *specialist* publishing services, Milton Keynes
Design and project management by Starfish DEPM, London.

Printed in Italy

Letts Educational Limited is a division of Granada Learning Limited, part of Granada plc.

Kinematics

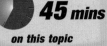
Spend no more than
45 mins
on this topic

Learn the key facts

1 An understanding of the following terms is essential when studying motion:

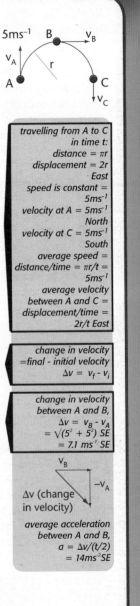

N ↑   5ms⁻¹ B v_B
v_A
A   C
r
v_C

- **Distance:** a scalar quantity equal to the total length travelled between two points. Units: metres (m).
- **Displacement:** a vector quantity equal to distance measured in a particular direction. It is the vector from the starting point to the finish point. Units: metres (m) in a given *direction* (e.g. right, +ve, East).
- **Speed:** a scalar quantity equal to the rate of change of distance. Units: metres per second (ms⁻¹)·

$$\text{speed} = \frac{\text{change in distance}}{\text{time for change}}$$

- **Velocity:** a vector quantity equal to the rate of change of displacement. Units: ms⁻¹ in a given direction.

$$\text{velocity} = \frac{\text{change in displacement}}{\text{time for change}}$$

Thus, an object may have constant speed (e.g. $5\,\text{ms}^{-1}$) but a changing velocity, as its direction may change (e.g. moving in a circle).

- **Acceleration:** a vector quantity equal to the rate of change of velocity. Its direction is that of the change in velocity. Units: ms⁻² in the direction of the change in velocity.

$$\text{acceleration} = \frac{\text{change in velocity}}{\text{time for change}} = \frac{\text{final velocity} - \text{initial velocity}}{\text{time for change}}$$

i.e. If a body's velocity changes, it accelerates.

The change in velocity is found using vector addition. Consider the example of moving in a circular path at constant speed. The object always changes its direction of travel in towards the centre of the circle. Therefore, it has an acceleration ($a = v^2/r$) in towards the centre of the circle.

> *travelling from A to C in time t:*
> *distance = πr*
> *displacement = 2r East*
> *speed is constant = 5ms⁻¹*
> *velocity at A = 5ms⁻¹ North*
> *velocity at C = 5ms⁻¹ South*
> *average speed = distance/time = $\pi r / t$ = 5ms⁻¹*
> *average velocity between A and C = displacement/time = 2r/t East*

> *change in velocity = final - initial velocity*
> *$\Delta v = v_f - v_i$*

> *change in velocity between A and B,*
> *$\Delta v = v_B - v_A$*
> *$= \sqrt{(5^2 + 5^2)}$ SE*
> *$= 7.1\ \text{ms}^{-1}$ SE*
>
> v_B
> $-v_A$
> Δv (change in velocity)

2 Linear motion may be represented graphically, with + and − values used to represent direction (e.g. 'up/down').

It is important to know the slope or gradient of a graph:

$$\text{gradient} = \frac{\text{change in vertical step}}{\text{corresponding change in horizontal step}}$$

> *average acceleration between A and B,*
> *$a = \Delta v/(t/2)$*
> *$= 14\text{ms}^{-2}$ SE*

Speed, velocity and acceleration may change, in which case we can define average and instantaneous values (see graphs).

The gradient of the dotted line on a displacement–time graph gives the *average velocity* between times A and C.

s/m   gradient = $\Delta s/\Delta t$
      = average velocity
s_C C
                  total displacement, Δs
A   total time
     Δt t/s

Kinematics

$$\text{average velocity} = \frac{\text{total displacement}}{\text{time taken}}$$

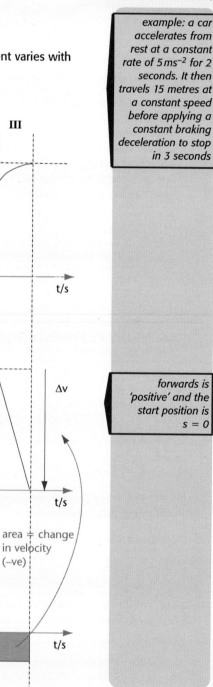

DAY
1
2
3
4
5
6
7

- A displacement–time (s–t) graph **shows how the vector displacement varies with time.**

The *gradient* at a point of the displacement-time graph gives the *instantaneous velocity* (i.e. at that point). Here the velocity, and thus gradient, (I) initially increases from zero up to a large and positive value, (II) then remains constant at the large, positive value before (III) decreasing to zero.

- A velocity–time graph (v–t) **shows how the vector velocity varies with time.**

The simplest way to sketch this graph is to plot how the gradient of the displacement–time graph varies with time (see description above). Note that the velocity is always positive here – i.e. the body always travels in the same direction.

The *area* under a section of the velocity-time graph gives the *change in displacement* over that time period. Thus, the total displacement is the total area under the graph.

Note that the area above the time axis is 'positive' and that below is 'negative'. If a body returns to its original position, the total displacement, and therefore area, is zero. In comparison the scalar quantity of '*total distance travelled*' is the total area under the graph irrespective of sign. To calculate the area, divide the graph into simple shapes (here two triangles and a rectangle).

The *gradient* at a point of the velocity–time graph gives the *instantaneous acceleration* (i.e. at that point). Here the acceleration, and thus gradient, (I) is initially constant and positive, (II) then is constant and zero, and finally (III) is constant and negative.

- An acceleration–time (a–t) graph **shows how the vector acceleration varies with time.**

example: a car accelerates from rest at a constant rate of $5\,ms^{-2}$ for 2 seconds. It then travels 15 metres at a constant speed before applying a constant braking deceleration to stop in 3 seconds

forwards is 'positive' and the start position is $s = 0$

Kinematics

The *area* under a section of the acceleration–time graph gives the *change in velocity* over that time period. N.B. A positive area shows an increase in velocity and a negative area, a decrease in velocity. In this example the increase in velocity (I) equals the decrease in velocity back to zero (III). Therefore, the areas above and below the time axis are equal and opposite.

- When dealing with motion questions:
  a) Always define which direction is taken as positive and where the zero displacement point is. *E.g. In the above example forwards is positive and displacement is zero at t = 0.*
  b) If the motion divides into distinct regions, treat them separately. *E.g. Regions I, II and III have different accelerations.*
  c) Draw a labelled diagram, showing all data given.

3 The equations of motion are used to solve problems concerning bodies with a *constant acceleration*.

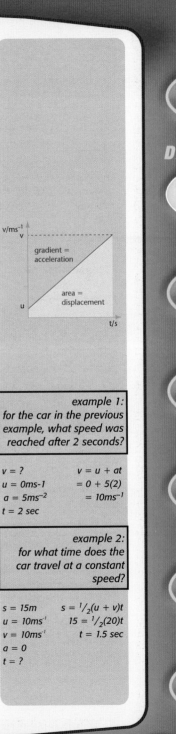

(1)	$v = u + at$	u = initial velocity (ms^{-1})
(2)	$s = \frac{1}{2}(u + v)t$	v = final velocity (ms^{-1})
(3)	$s = ut + \frac{1}{2}at^2$	a = acceleration (ms^{-2})
(4)	$v^2 = u^2 + 2as$	t = time (s)

- Only use these equations for time intervals, or 'regions', when the acceleration has a constant value *(e.g. a different set of values are required for each 'region' in the above example).*
- For each region, write down values of *s, v, u, a,* and *t,* using any information given in the question (e.g. 'highest point' means *v=0,* 'original position' means *s=0*). Ensure all values have correct units and signs (for direction).
- Select the appropriate equation of motion and solve it.
- Any body which has a constant resultant force acting on it will have uniform acceleration, e.g. a falling object in the absence of drag forces, when only gravity acts (i.e. in free fall).

4 • Non-uniform acceleration

If acceleration is non-uniform, the 'equations of motion' CANNOT be used. A body which has a changing resultant force acting on it will have non-uniform acceleration.

example 1:
for the car in the previous example, what speed was reached after 2 seconds?

$v = ?$ $v = u + at$
$u = 0ms\text{-}1$ $= 0 + 5(2)$
$a = 5ms^{-2}$ $= 10ms^{-1}$
$t = 2$ sec

example 2:
for what time does the car travel at a constant speed?

$s = 15m$ $s = \frac{1}{2}(u + v)t$
$u = 10ms^{-1}$ $15 = \frac{1}{2}(20)t$
$v = 10ms^{-1}$ $t = 1.5$ sec
$a = 0$
$t = ?$

Kinematics

Example 1: A falling object where drag acts as well as gravity, leading to terminal velocity, v_t (see 'Force and Momentum').
Due to the drag force increasing with velocity, the total force acting on the object gets smaller, until it reaches zero, as the velocity increases.

Example 2: A railway-car hitting a buffer where a spring is compressed.

example 2:
a possible a–t graph for a spring being put under compression and then released

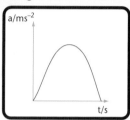

example 1:
terminal velocity:
v–t and a–t graphs

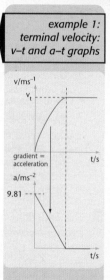

In this case the force and hence acceleration varies according to how much the spring is compressed or stretched.

• 2-D projectile

A particle may not simply move along a straight line (i.e. up/down or forwards/backwards), but may move in two dimensions (i.e. horizontally as well as vertically). In this case, *the horizontal and vertical motions are considered separately as they are independent of one another.*

For a particle in flight (neglecting drag/air resistance), a constant acceleration acts vertically down (due to gravity) and, as no force acts horizontally, it has constant horizontal velocity. Thus:

Vertically: constant acceleration = g use 'equations of motion'
Horizontally: constant velocity = speed use speed = distance/time

N.B. The time of flight is the same whether considering the horizontal or vertical motion.

Example: A marble rolls off a 1m high table at $2\,ms^{-1}$. When does it hit the ground? How far has it travelled horizontally?

• VERTICALLY: let down be positive and s = 0 at table top.
  At the ground $s = 1m$; $u = 0\,ms^{-1}$; $a = 9.81\,ms^{-2}$; $t = ?$
  $s = ut + \frac{1}{2}at^2$; $1 = 0 + \frac{1}{2}(9.81)t^2$; $t = +0.45$ seconds

• HORIZONTALLY: speed $= 2\,ms^{-1}$; time = 0.45s;
  distance $= vt = 2(0.45) = 0.90m$

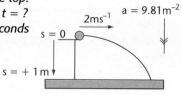

Have you improved?

1 During a game of squash, a ball takes 1.7 s to travel from player **A** to player **B**, along the path shown in the diagram.

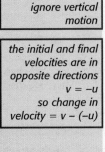

2.0 m

1.4 m

A

1.5 m

3.0 m

B

1.6 m

6.0 m

a) Calculate:
   i) the total distance that the ball travels
   ii) the final displacement
   iii) the average speed
   iv) the average velocity.

b) The ball hits a racquet with a speed 8 ms^{-1}, and leaves in the opposite direction at 14 ms^{-1}. Calculate the acceleration of the ball if it is in contact with the racquet for 0.06 s.

ignore vertical motion

the initial and final velocities are in opposite directions
$v = -u$
so change in velocity = $v - (-u)$

2 a) Sketch the (i) displacement–time; (ii) velocity–time and (iii) acceleration–time graphs for a bouncing ball released from rest 1.25 m above a horizontal surface. Explain your answer fully.

b) i) With what velocity does the ball hit the ground?
   ii) If the ball rebounds with half this velocity, what height does it rebound to?
   iii) What time elapses between the first and second bounce?

ignore air resistance.
take g = 10 ms^{-2}

take upward as positive and ground level as s = 0

3 a) A ball is projected horizontally from a table at a speed of 10 ms^{-1}.
   i) How far below the table-top will the ball be after 0.4 seconds?
   ii) How far will it have travelled horizontally at this time?

b) What will its (i) horizontal and (ii) vertical velocity be at this time? (iii) What will its resultant velocity be at this time?

ignore air resistance
take g = 10 ms^{-2}

take upward as positive and the table top as s = 0

4 a) A body's velocity varies with time as shown. If the body is initially travels to the right, in which way does it move at (i) $t = 1.5$ s and (ii) 3 s?

b) i) Which feature of the graph shows the change in displacement of the body between $t = 0$ to $t = 1.5$ s?
   ii) What can be said about the total displacement at $t = 3$ s?

c) i) How can the acceleration of the body be determined from the graph?
   ii) Sketch an acceleration–time graph for this motion.
   iii) What does the total area under this graph give?

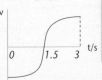

v

0   1.5   3   t/s

Force and Momentum

How much do you know?

1 a) The momentum of a body is the product of its _____ and _____.
   b) A cannon of mass 700 kg fires a cannon ball of mass 50 kg horizontally with a velocity of 190 ms^{-1}. Calculate the recoil velocity of the cannon.

2 a) Force is a _____ quantity with the unit _____. A resultant force can cause an object to change its velocity, i.e. it _____.
   b) List four different types of force.

3 a) Draw a free-body diagram for a plane climbing at a steady rate and a constant speed.
   b) What is the resultant force? Explain your answer.

4 Newton's first law states that a body which has no external forces acting on it will remain or continue to move in a _____ _____ with zero _____.

5 a) If a footballer kicks a stationary ball with a force of 100N, and it acquires a momentum of 5 kgms^{-1}, how long is the foot in contact with the ball?
   b) What impulse does he give the ball?

6 A ball of mass 1.2 kg falls towards the Earth under the action of gravity. If it has an acceleration of 6.5 ms^{-2}, determine the magnitude and direction of the air resistance acting on the ball. Take g = 9.81 ms^{-2}.

7 a) A lamp hangs from a ceiling. Draw a free-body diagram for the lamp and identify all the Newton pair forces to those in your diagram.
   b) A spring and two masses are in equilibrium as shown. The thread joining the masses is cut. Calculate the magnitude and direction of the accelerations on both masses the instant the thread is cut.

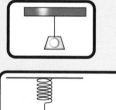

m = 0.6 kg

m = 0.2 kg

Answers

1 a) mass / velocity b) −13.6ms^{-1} **2** a) vector / Newton / accelerates
b) any 4: weight, electrical, magnetic, upthrust, drag, normal
reaction, frictional, tension **3** b) zero resultant force as there is
no acceleration vertically or horizontally **4** stationary / straight
line / acceleration **5** a) 0.05 s b) +5Ns **6** 4.0N up **7** b) 3.3ms^{-2}
upwards, 9.8ms^{-2} downwards

7 a)

tension in cord on lamp
N3: pull of lamp down
on cord

Weight N3: pull of lamp
up on Earth

3 a)

lift

thrust

drag

weight

If you got them all right, skip to page 23

Learn the key facts

1 Momentum is defined as the product of a body's mass and velocity.

> momentum = mass × velocity   ($p = mv$)

- Momentum is a vector quantity with the units kgms^{-1} or Ns.
- Momentum is important as it is related to force (see later).
- Momentum is conserved.

The Principle of Conservation of Momentum states that if no resultant external forces (e.g. friction) act upon a system of interacting bodies then the total momentum is constant.

- *Inelastic collision (momentum conserved)*

momentum before collision = momentum after
$$m_1u_1 + m_2u_2 = (m_1 + m_2)v$$

Explosion

momentum before spring release = momentum after
$$0 = m_1v_1 + m_2v_2$$
$$m_1v_1 = -m_2v_2$$

The minus sign indicates opposite directions.

- *Elastic collision (momentum and kinetic energy conserved)*

Momentum before collision = momentum after
$$m_1u_1 + m_2u_2 = m_1v_1 + m_2v_2$$

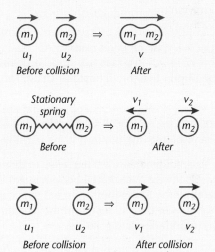

Kinetic energy before collision = kinetic energy after
$$\tfrac{1}{2}m_1u_1^2 + \tfrac{1}{2}m_2u_2^2 = \tfrac{1}{2}m_1v_1^2 + \tfrac{1}{2}m_2v_2^2$$

- Any collision in which the total kinetic energy is not constant, is termed an **'inelastic' collision**.
- As momentum is a vector, define a positive direction and keep to it throughout the question.
- Always draw a 'before' and 'after' collision diagram.

2 A force is a vector quantity and has the unit newton (N). Forces can be considered as pushes or pulls. Forces can cause objects to accelerate, decelerate, rotate, change direction, change shape, change size, or break.

There are several types of forces:

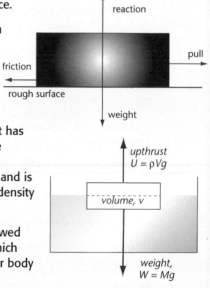

$W = mg$
mass
acceleration due to gravity ($9.81ms^{-2}$ on Earth)

- Weight is the gravitational pull of the Earth (Moon, etc.) on a body.
- Contact forces occur when two bodies are touching. They are of two types:

The normal contact (or reaction) force acts at 90° to the common surface.

The friction force acts along the common surface, to oppose any motion between the two bodies. 'Rough' surfaces have friction, whereas 'smooth' surface are taken to be frictionless.

reaction

friction

pull

rough surface

weight

- Drag forces act to slow bodies down that are moving through a fluid (e.g. liquids, air). They increase in magnitude with increasing speed. E.g. air or water resistance.
- Tension exists in a taut (i.e. stretched) wire, spring, string or cable. It has the same value throughout the string. The wire becomes slack if the tension is removed (i.e. if the tension becomes zero).
- Upthrust acts upwards on objects that are in a fluid (e.g. air, liquid) and is equal to the weight of fluid displaced. If a volume V of a fluid with density ρ is displaced, the upthrust is $mg = (\rho V)g$.

upthrust
$U = \rho Vg$

volume, v

weight,
$W = Mg$

3 For mechanics problems it often helps to draw a situation diagram followed by a free-body diagram. A free-body diagram is a simplified diagram, which illustrates all the forces (both size and direction) acting on one particular body only. Arrows are used to represent the direction of the forces, and their lengths indicate their magnitudes.

The resultant force, F_R, acting on a body can be found by adding all the forces acting on an object (i.e. all the forces on the free-body diagram). If the forces acting on a body are not parallel, they can be combined (see 'Vectors'):

the above diagrams are free-body diagrams

- diagrammatically with a polygon of forces, *or*
- by resolving into perpendicular components (e.g. horizontal and vertical components) and then treating the components separately.

the resultant forces on the floating object:
horizontally: $F_R = 0$
vertically: $U - W = F_R$

4 Newton's First Law of motion states: If no external resultant force acts upon a body, it will either remain stationary or continue to travel in a straight line with constant velocity.

- i.e. balanced forces don't alter the velocity of an object

DAY

2

Example 1: A stationary floating object of mass M (see p 20)
Resolve forces vertically: $U - W = 0$; $U = W$

Example 2: A plane with constant velocity (i.e. steady speed at a constant altitude).
Resolve horizontally: Thrust (T) = Drag (D)
Resolve vertically: Lift (L) = Weight (W)
N.B. The lift always acts at 90° to the wings.

Example 3: Forces at angles
A skier of weight 500N is towed along a rough, flat piece of ground at a constant speed by an overhead tow-rope as shown. Calculate T.
The resultant force is zero, both vertically and horizontally:
Vertically:　　　$R + T\sin30 - 500 = 0$
Horizontally:　　$T\cos30 - 13 = 0 \Rightarrow$　　$T = 15N$

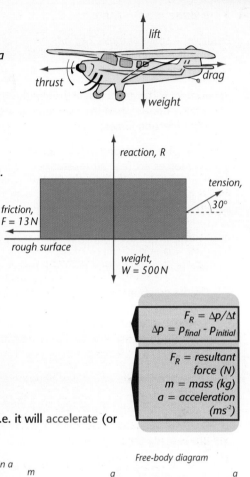

Newton's Second Law of motion **states**: The rate of change of momentum of a body is equal to the resultant force acting on the body, in the direction of that force.

Resultant force (F_R) = $\dfrac{\text{change in momentum } (\Delta p)}{\text{time taken } (\Delta t)}$

$= \dfrac{\text{final} - \text{initial momentum}}{\text{time taken}}$

$F_R = \Delta p / \Delta t$
$\Delta p = P_{final} - P_{initial}$

F_R = resultant force (N)
m = mass (kg)
a = acceleration (ms^{-2})

- For a body of constant mass, this becomes:

'If a resultant force acts upon a body it will change its velocity', i.e. it will accelerate (or decelerate) where:

$F_R = ma$

Example:
car:　resultant force = $F - T$
so:　　$F - T = Ma$
caravan:　resultant force = T
so:　　$T = ma$

both have acceleration a

F: engine force

T
tension in towbar

Free-body diagram

Car

Caravan

- Remember that a body with constant speed can still accelerate, if it changes its direction of travel – as velocity is a vector quantity which therefore changes with direction. E.g. a satellite orbiting the earth at constant speed and altitude. In this case, the satellite's weight provides the 'accelerating' force 'inwards'.
- The Newton *is defined as the unit of force so that a force of 1N accelerates a mass of 1kg at 1ms⁻².*

$F = ma$
$a = v^2/r$

- Impulse *is defined as the total change in momentum, Δp, of a body. If a constant, or average, force F is applied over a time Δt;*

$Impulse = \Delta p = F\Delta t$

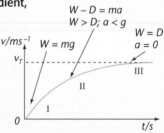
impulse = change of momentum

Example: A stake of mass 2 kg hits the ground at 5 ms^{-1} and takes 2 s to come to rest.
$Impulse = \Delta p = p_{final} - p_{initial} = 0 - 2(5) = -10 Ns$
$Average\ force = I/\Delta t = -\frac{10}{2} = -5N$, i.e. a retarding force of 5 N.

6 Application of Newton's first and second laws

I Free fall – occurs when the only force acting is the body's weight (i.e. there are no drag or friction forces). Thus, the acceleration is constant and equals 9.81 ms^{-2} (i.e. 'g').

II If drag forces act (e.g. air resistance), then the acceleration decreases as the body speeds up. This is due to the resultant force on the body decreasing as the drag force increases with increasing speed:
$W - D = ma$

III At some speed the drag force will exactly balance the weight, leaving no resultant force: $W - D = 0$.
As there is no acceleration, the body travels at a constant, or terminal velocity.

- On the graph of velocity against time, the acceleration is shown by the gradient, which decreases from large and positive to zero (see kinematics).
- Similarly, a car with a constant thrust or driving force reaches a maximum velocity, due to drag increasing with speed.

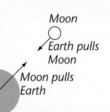
freefall
weight
$w = mg$

$W - D = ma$
$W > D; a < g$
$W = D$
$a = 0$
v/ms^{-1} $W = mg$
v_T
III
II
I
0
t/s

7 Newton's Third Law of motion states: If a body A exerts a force on body B, then body B exerts an equal and opposite force on body A.

A pair of forces that obey this law are called a Newton pair of forces.
Specifically they:

i)   are equal in size
ii) are opposite in direction
iii) act on different bodies
iv) act along the same line
v)   are the same type of force (e.g. gravitational).

Moon
Earth pulls Moon
Moon pulls Earth
Earth

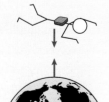
Earth pulls skydiver downward

Skydiver pulls earth upwards

Car pushes road backward = Road pushes car forwards

Have you improved?

1 A 2.0 kg object moving with a velocity of 10.0 ms^{-1} collides with a 4.0 kg object moving with a velocity of 5.0 ms^{-1} along the same line. If the objects stick together on impact, calculate their common velocity when initially moving a) in the same direction, and b) in opposite directions.

2 A rocket engine ejects 100 kg of exhaust gas over 5 seconds at a velocity of 250 ms^{-1} relative to the rocket.
a) Calculate the change in momentum of the rocket.
b) Calculate the average force on the rocket.

> *assume 100 kg is negligable in comparison to the mass of the rocket*

3 Draw free-body diagrams for a hot-air balloon: a) just before take-off; b) while hovering at constant height just after take-off; c) when it accelerates upwards. For each case identify all the forces acting, and the direction of the resultant force.

4 If the resistive forces acting on the skier and the boat are 80 N and 200 N respectively, determine:
a) the tension in the rope
b) the driving force of the speed boat engine.

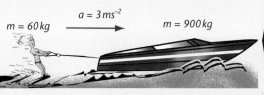

> *draw free-body diagrams for (a) the skier and (b) the boat*

5 The variation of the forces acting on a skydiver of mass 50 kg with his speed are shown in the graph.
a) Draw a free-body diagram of the skydiver.
b) What on the force–velocity graph represents the resultant force at a given velocity?
c) Calculate the resultant force at 0 ms^{-1}, 20 ms^{-1} and 45 ms^{-1}.
d) Determine the terminal velocity.

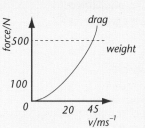

6 A book rests on a table, which is on the Earth as shown. Draw separate free-body diagrams for the:
a) book,
b) table and
c) Earth.

Equilibrium and Moments

How much do you know?

1 a) A turning force is called a _____ or a _____ and is equal to the product of the size of the _____ and the _____ distance of the force from the _____ . It has units of ____.

b) Determine the magnitude and direction of the turning forces produced in the diagrams.

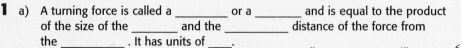

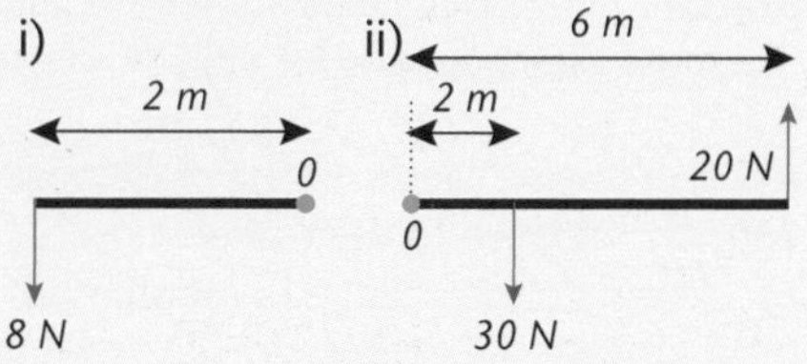

i)
2 m

8 N

ii)
6 m
2 m
20 N
0
30 N

2 a) A couple consists of _____ forces of _____ magnitudes acting in _____ directions, causing an object to _____.

b) Determine the magnitude and direction of the couple produced in the diagram opposite.

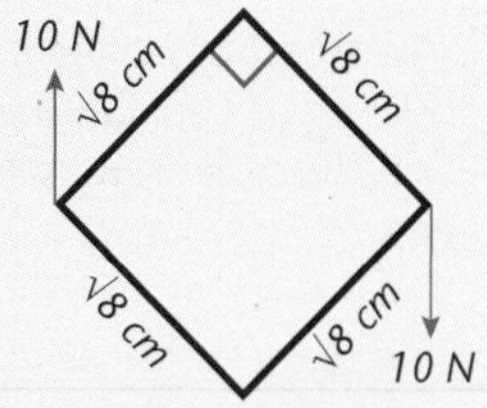

10 N
√8 cm
√8 cm
√8 cm
√8 cm
10 N

3 If the system of forces shown in the diagram is in equilibrium, determine the magnitude of *F*.

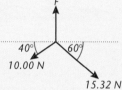

F

40°   60°
10.00 N
15.32 N

4 If the see-saw is in equilibrium, determine the mass, *m*.

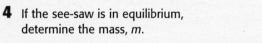

0.8 m   0.4 m   1 m

8 kg   m   1 kg

Answers

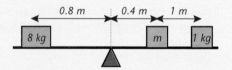

If you got them all right, skip to page 27

Equilibrium and Moments

Spend no more than
45 mins
on this topic

Learn the key facts

1 Application of a force may cause a body to rotate. This turning force is called a moment or torque. The size of the moment depends upon the size of the force and the separation between its point of application and the turning point (pivot / fulcrum), i.e.

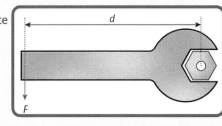

$T = F \times d$ (anti-clockwise here)
T = torque/moment (Nm)
F = force (N)
d = perpendicular (i.e. shortest) distance between force and pivot (m)

2 Two forces produce a couple if they act in opposite directions with equal magnitudes, causing an object to rotate.

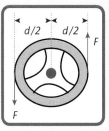

e.g. turning a steering wheel; total turning force (couple)
$= F \times \frac{d}{2} + F \times \frac{d}{2}$
$= Fd$ (clockwise here)
d = perpendicular separation of forces

3 The two conditions for equilibrium of co-planar forces are:

1) Total forces = 0
2) Total moments = 0

Any system of forces in equilibrium must satisfy both of these requirements.

- The first condition for equilibrium is that **no resultant force acts upon the body, in any direction.** This is Newton's First Law (see 'Force and Momentum'). If a body remains stationary, or moves with a fixed velocity under the action of a number of forces, then no resultant force acts upon the body, i.e. the forces 'cancel each other'.

E.g. If the supporting string holds the mass stationary, the system must be in equilibrium. The vertical components of the tension in the strings, must balance the weight of the body.

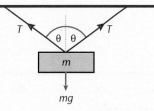

$T\cos\theta + T\cos\theta = mg$;
$2\,T\cos\theta = mg$

- The second condition for equilibrium is that there should be *no resultant turning force about any axis.*

25

Equilibrium and Moments

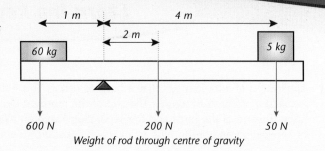

The principle of moments states: *If a body is in equilibrium, the total clockwise moment about any axis is equal to the total anti-clockwise moment about that axis.*

e.g. anti-clockwise moment
= 600 x 1 = 600 Nm

clockwise moment
= (2 x 200) + (4 x 50) = 600 Nm

clockwise moment = anti-clockwise moment, so system is in equilibrium (balances).

DAY **4** a) If there is more than one unknown force, take moments about the point through which one of them acts – usually the point through which the force(s) that you do not want to know anything about.

Example

Find the force, F, required to support a uniform car bonnet of length d and weight W that pivots about a point O.

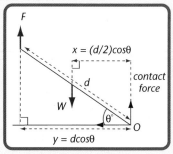

- Draw a fully labelled diagram, showing all forces, lengths and angles.
- The centre of gravity of a body is the point at which all the weight of the body can be considered to act. If it lies inside the body, it is the point at which the body 'balances'. If the body is 'uniform', then this is at the object's centre of symmetry. E.g. here it is at the centre of the car bonnet.
- Take moments about O, to exclude the contact forces at O.
- Apply total moments = 0, remembering to use the perpendicular distance from the axis to the force.
  $Wx = Fy$; $F = Wx/y$

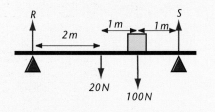

b) Both conditions for equilibrium may be needed to solve a problem.

Example

A uniform beam of weight 20N is supported by two tressels as shown. A box of weight 100N rests 1m from one tressel. Find the reaction forces on the beam due to each tressel.

- More than one unknown force, so take moments about one of them, say R:
  $2(20) + 3(100) = 4S \Rightarrow S = 85N$
- Resolve forces vertically (here none act horizontally):
  $R + 85 = 20 + 100 \Rightarrow R = 35N$

Have you improved?

1 The body in the diagram has a weight of 800 N and is stationary due to the frictional force between the surface and the box. Determine the magnitude of this frictional force.

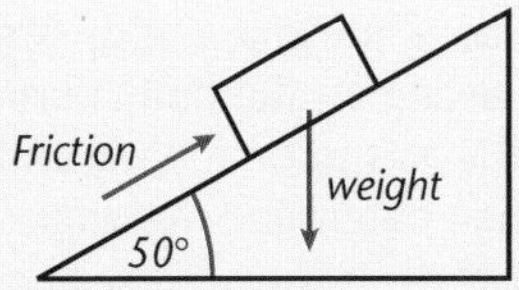

Friction *weight* 50°

> the component of the weight acting down the slope ($w\cos 40°$) = frictional force

2 The system shown in the diagram is in equilibrium and the maximum tension that the supporting string can withstand before snapping is 100 N.

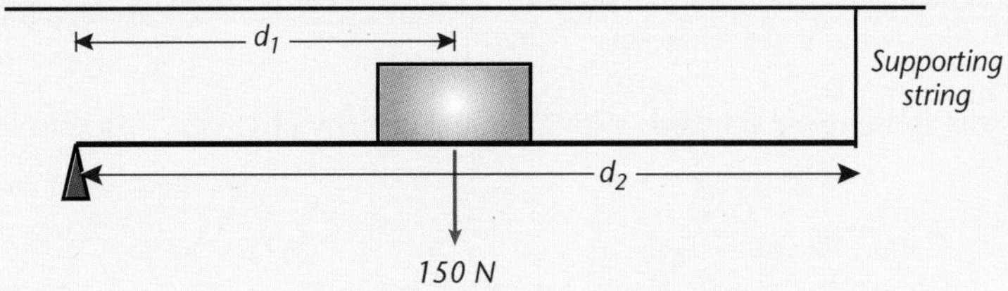

d_1 *Supporting string* d_2 *150 N*

> use Principle of Moments

Determine the minimum possible value of the ratio d_2/d_1

3 The body, p, shown in the diagram, has a mass of 2 kg and travels to the left with a steady speed of 6 ms^{-1}.
a) Determine the angle θ.
b) If the body is to accelerate towards the left at 5 ms^{-2}, with the 8 N force and θ unchanged, determine the magnitude of the other pair of forces, if they are to have magnitudes equal to each other.

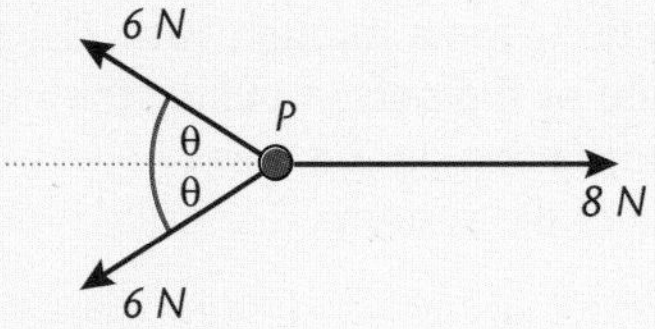

6 N P θ θ 8 N 6 N

> sum of components of 6N forces to the left must equal 8N, steady speed means system is in equilibrium

> use $F = ma$ to find resultant force to the left (10N)
> $2F\cos\theta = 10 + 8$

4 A uniform plank of mass 5 kg rests on two supports, R and S, with two 10 kg boxes placed as shown. Take $g = 9.8$ ms^{-2}.

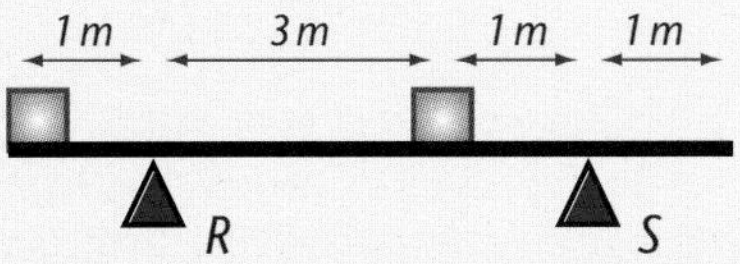

1 m   3 m   1 m   1 m
R   S

> 'rests' means the plank is in equilibrium

> take moments about R or S

a) Label the forces acting on the plank.
b) Calculate the value of the reactions at the supports.

DAY 3

How much do you know?

1 a) 60 J of work is done moving a body 8 m along a frictionless horizontal surface. Determine the resultant force acting on the body during its movement.
   b) A force of 1000 N acts upon a body and causes it to move 400 m. If the work done on the body is 30 kJ then determine the angle between the force and the direction of the motion.

2 a) A β-particle can have a velocity up to 98% of the speed of light. Calculate the kinetic energy of the fastest β-particle, ignoring relativistic effects.
   b) A crate of mass 60 kg is lifted onto a shelf 3 m high. Calculate the potential energy gained by the crate. (Take $g = 10\text{ms}^{-2}$.)

speed of light = $3.0 \times 10^8\text{ms}^{-1}$

mass of electron = $9.1 \times 10^{-31}\text{kg}$

3 A ball of mass 0.6 kg is thrown directly upwards with an initial velocity of 15 ms^{-1}. Determine its:
   a) initial KE
   b) maximum height.

4 A child of mass 35kg runs up a flight of 20 stairs in 1.8 s. Each stair has a vertical rise of 0.15 m. Determine the average power developed by the child. (Take $g = 10\text{ms}^{-2}$.)

5 Using a double pulley system to lift an 80 kg load, a pulling force of 420 N is moved through 20 m in order to lift an 80 kg mass through 10 m. Calculate:
   a) The work done by the person pulling the rope.
   b) The useful work done on the mass.
   c) The efficiency of the system.

Answers

1 a) 7.5 N b) 85.7° **2** a) 3.9 × 10⁻¹⁴J b) 1800 J **3** a) 67.5 J b) 11.3 m **4** 583 W
5 a) 8400 J b) 8000 J c) 95%

If you got them all right, skip to page 32

Learn the key facts

1 If a body has energy, it has the potential to do work. Doing work could be lifting a box, pushing a car or a light bulb converting electrical energy into light energy, for example. The more energy that a body has, the more work it can potentially do. Both work and energy have the unit joule (J).

- If doing work involves the movement of a force then:

  $W = Fs$

  W = work (J)
  F = force (N)
  s = distance moved in direction of force (m)
- If the movement is in a direction other than that in which the force is applied then:

  $W = Fs \cos\theta$
- If the force varies, then take the average force as the applied force, F.

Work is often defined as: *the amount of energy converted from one type to another.*

2 Different energy types can be broadly categorised into two main groups: kinetic energy and potential energy.

- Kinetic energy is the energy a body possesses by virtue of its motion. The faster the body moves, the greater its KE.

  $KE = \frac{1}{2}mv^2$ m = mass of body (kg)
  v = speed (ms^{-1})
- Potential energy is energy that is stored in some manner. Often this is energy that is possessed by virtue of configuration or position, e.g. the energy stored in a stretched spring or elastic, called the elastic potential energy, or the energy a body has due to its position in a gravitational field, called gravitational potential energy.

Elastic potential energy

$EPE = \frac{1}{2}F\Delta x = \frac{1}{2}k(\Delta x)^2$

F = maximum force (N)
Δx = spring extension (m)
k = spring constant (Nm^{-1})

Example: An initially unstretched spring is stretched to have an extension of 10cm. What force is exerted at this point if the spring stores 100J?

$EPE = 100 = \frac{1}{2}F(0.10); F = 2kN$

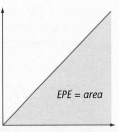

force
F/N

EPE = area

extension, Δx/m

Work, Energy and Power

Gravitational potential energy

$$GPE = mgh$$

m = mass of body (kg)
g = acceleration due to gravity (ms^{-2})
h = height (m) above a zero GPE reference point.

This is only valid when near the surface of the Earth.

Example: A boy of mass 45 kg runs a 5 km race which starts at 300 m above sea level and climbs steadily to finish at 527m. What is the change in GPE?

Only the change in height will change the GPE;
$GPE = 45(9.8)(527 - 300) = 100 kJ$

DAY

3

3 The Principle of Conservation of Energy states: ***in an isolated system (e.g. the Universe) the total amount of energy does not change.***

This statement implies that energy can neither be created nor destroyed. It can, however, be converted from one type to another.

- If an isolated system consists of bodies with KE and PE, and no mechanism by which these can gain other forms of energy, then there will be an interchange of PE and KE, but the total energy will remain constant, i.e.

$$KE + PE = constant$$

$\frac{1}{2}mv^2 + mgh$ = constant

If a mass fell towards Earth under the influence of gravity alone, then all of the PE it had at height, h, would be converted to KE at the Earth's surface, i.e.

$\frac{1}{2}mv^2 = mg\Delta h$
$v = \sqrt{(2g\Delta h)}$

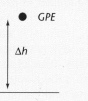

- Systems that are not isolated allow energy to be transferred either out of or into the system. Dissipative forces (e.g. friction) cause energy to be transferred out of a system.
This is summarised in the Work–Energy Principle:

work done = change in system energy

$$Fs = \Delta(KE + PE)$$

Where F = total external force (N)
s = distance moved in direction of force, F (m)

Example: A car's engine stops while it is travelling on a flat surface at $20\,ms^{-1}$. If the resistive forces on the car, of mass $1000\,kg$, are $200\,N$, how far will the car travel before coming to rest?

Initial energy of the car $= KE = \frac{1}{2}(1000)20^2 = 200\,kJ$

$\Rightarrow 200{,}000 = 200s; \; s = 1000m$

4 The rate at which work is done, i.e. the rate at which energy is converted from one form to another, is known as power.

$$\text{power} = \frac{\text{work done}}{\text{time taken}} = \frac{\text{energy converted}}{\text{time taken}}$$

So:

$$P = \frac{Fs}{t} \qquad \text{but } \frac{s}{t} = \text{speed } (v) \quad \Rightarrow \quad P = Fv$$

Power has the units joules per second (Js^{-1}) or watts (w).

5 Although energy cannot be destroyed, it can be converted into forms of energy that are not useful. Machines and engines often 'lose' energy as heat and sound, due to friction between moving parts. A machine that 'loses' a large fraction of its input energy is said to have a low efficiency, where:

$$\text{efficiency \%} = \frac{\text{useful energy out}}{\text{total energy in}} \times 100$$

or

$$\text{efficiency \%} = \frac{\text{useful power out}}{\text{total power in}} \times 100$$

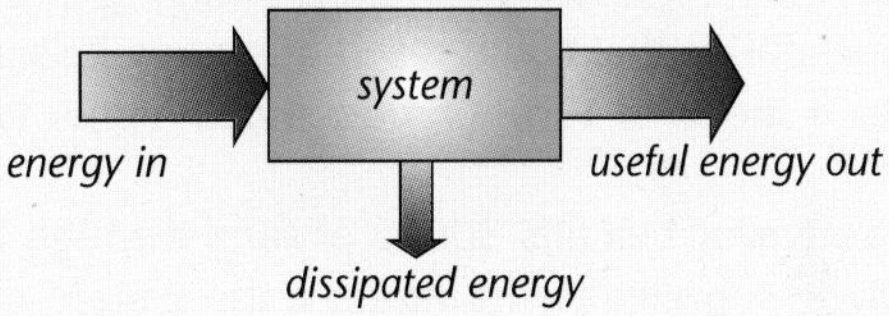

energy in → system → useful energy out

dissipated energy

Work, Energy and Power

Have you improved?

DAY

3

1 A cyclist approaches a slope with a speed of $7\,ms^{-1}$. At the base of the slope the cyclist stops pedalling and freewheels up the slope. Calculate the distance, s, travelled up the slope if the percentage resistive energy losses are:

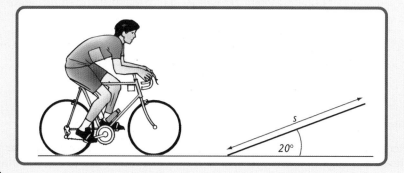

a) 0%
b) 50%
c) 70%

$g = 10ms^{-2}$

$kE\ bottom = PE$ top

2 Consider a 'perfect' pendulum swinging in the absence of friction or any other mechanism that could result in energy losses. If at the lowest point of its swing, it has a speed of $10\,ms^{-1}$, how much higher would it be at the top of its swing?

$g = 10ms^{-2}$

3 A conveyor belt lifts 300 kg of sand through a height of 10 m, in 15 s. Calculate the power supplied to the motor if the system is:
a) 100% efficient
b) 24% efficient
c) 50% efficient.

$g = 10ms^{-2}$

$w = mgh$
$P = {}^w/_t$

4 a) The engine of a car travelling at a steady speed of $72\,kmh^{-1}$, on a level road, operates at 100 kW. Determine the magnitude of the forces that resist this motion.
b) If the car had a mass of 900 kg, and were to drive up a slope inclined at 30° to the horizontal, how much extra power would the engine have to develop to maintain the car's speed at $72\,kmh^{-1}$?

rearrange:
$P = Fv$

extra power = PE gained each second, so find height gained each second

5 A ball of mass 0.25 kg is thrown vertically upwards with a velocity of $15\,ms^{-1}$. If it reaches a height of 10 m, calculate
a) The percentage transformation in energy caused by air resistance.
b) The average resistive force.

$g = 10ms^{-2}$

How much do you know?

1 A DC supply is placed across the ends of a wire conductor, causing 500 C of charge to flow past a point in the wire, in 1000 s.
   a) Calculate the current in the wire.
   b) If the current were halved, calculate the time for 500 C to pass the point.

2 The diagram shows a section of wire that has a potential difference applied across its ends. Add to the diagram clearly labelled arrows to show the typical path of electrons, the direction of net charge movement, and the conventional current direction.

3 A charged particle of 2 x 10^{-10} C moves from one point to another, losing 2 x 10^{-8} J of electrical energy.
   a) Determine the PD between the points.
   b) If 10 x 10^6 of these particles move between the points in 40 s, determine the current flowing.

4 If a conductor obeys _____ law, its current must be _____ to the _____ _____ across its ends. This relies upon its temperature being _____.

5 Sketch a graph of resistance against temperature for a typical semiconductor.

6 For a wire that carries a current of 0.2 A and has a resistance of 2 Ω, determine the total energy dissipated in 10 minutes.

7 Calculate the resistivity for a wire conductor of length 2 m with a radius of cross-section 2 mm and resistance 5 Ω.

Answers

5

resistance Ω

R₀ -----

semiconductor α < 0

temperature °C

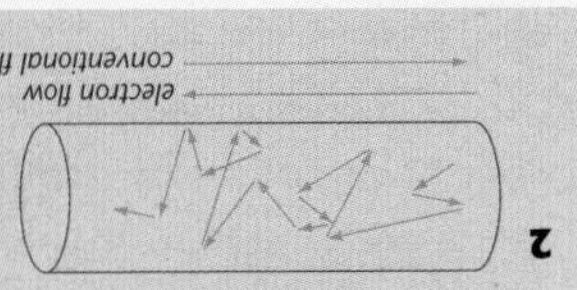

electron flow
conventional flow

2

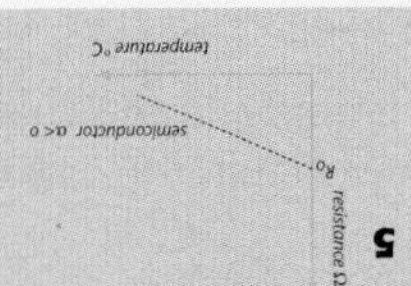

1 a) 0.5 A b) 2000 s **2** see below **3** a) 100 V b) 5x10⁻⁵A **4** Ohm's/proportional/ potential difference/constant **5** see below **6** 48 J **7** 3.1x10⁻⁵ Ωm

If you got them all right, skip to page 38

Current Electricity

Learn the key facts

1 Current (symbol I : unit amperes [A]) is a flow of charged particles, these are electrons in metals.

Definition: *the number of coulombs of charge passing a point in a circuit every second.*

i.e.

$$I = \frac{\Delta Q}{\Delta t}$$

This equation is used to define the unit of charge, the coulomb (C).

$$\Delta Q = I \, \Delta t$$

A charge of one coulomb passes at point when a steady current of one ampere flows for one second.

2 Electron flow is the direction of net electron movement through a component.

Conventional current (as labelled on circuit diagrams) is taken to be in the opposite direction to electron flow.

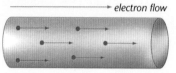

electron flow

conventional current

In a metallic conductor the charge carriers are free electrons. When a current flows in this wire the charge carriers move with a great speed ($\approx 10^6 \text{ms}^{-1}$). They do not, however, move through the material in a straight line; they are scattered randomly by the atoms of the material. If their speed were calculated by measuring the time they took to travel through a conductor of known length, a much smaller value would be found ($\approx 1 \text{mms}^{-1}$) due to the tortuous route they take through the conductor. This is known as the drift velocity.

speed $\approx 10^6 \text{ms}^{-1}$

drift velocity 1mms^{-1}

Consider the case where a current I passes through a wire of cross-sectional area A, length L, with a drift velocity v. Let n = the number of charge carriers per unit volume and q the charge carried by each charge carrier.

The product of the number of charge carriers per unit volume and the volume gives the number of charge carriers in the sample (N).

$$N = nLA$$

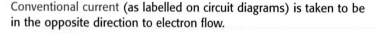

L

A

The product of the number of charge carriers and the charge on each carrier gives the total number of coulombs of charge contained within the sample.

$$\Delta Q = Nq = nLAq$$

This can be substituted into the current defining equation;

$$I = \frac{\Delta Q}{\Delta t} = \frac{nLAq}{\Delta t}$$

Drift velocity $(v) = L/\Delta t$, $\therefore$ $\boxed{I = nAqv}$ ($I = nAev$ for electrons)

3 A potential difference (symbol *V*: unit volts [V]) results in charge flow. Charge will only flow between two points if they are at different potentials.

Definition: *the potential difference between two points is the amount of electrical energy (W) converted into other forms per coulomb of charge passing between the points.*

$$V = \frac{W}{Q}$$

This equation can be used to define the volt i.e. *a one volt potential difference between two points will mean one coulomb of charge would dissipate one joule of energy when moving from one of the points to the other.*

4 · All materials (apart from superconductors), to some extent, hinder the passage of current through them. Resistance is a measure of the degree to which a component 'resists' current.

Resistance is defined as the ratio of the potential difference across a component to the current through it and has the unit ohm (Ω).

$$R = \frac{V}{I}$$

This equation can be used to define the ohm, i.e., a component has a resistance of one ohm if a potential difference of one volt across it causes a current of one ampere.

Certain conductors have a resistance that is constant for all values of *V* and *I*. We can see from the previous equation that for such a conductor *V* α *I*, i.e. Ohm's Law is obeyed (the material is Ohmic).

Ohm's Law describes the relationship between current and voltage and is stated as: *the current through a conductor is proportional to the potential difference across its ends as long as all other physical conditions remain constant.*

· Not all components respond in the same manner to different voltages and currents. For instance, when metals get hot their atoms vibrate increasingly, making the passage of electrons through the material more difficult, thus increasing its resistance. The materials behaviour can be shown graphically as an *I–V* plot using the circuit illustrated.

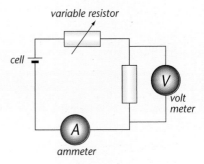

The graphs overleaf show some of the more common *I–V* plots.

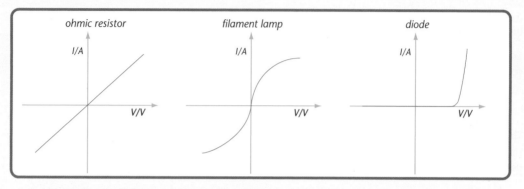

5 · A material's resistance is often dependent upon temperature and can be defined by the following equation:

$$R = R_0(1 + \alpha\theta)$$

R = resistance (Ω)
R_0 = resistance at 0°C
α = temperature coefficient of resistance
θ = temperature (°C)

For metals $\alpha > 0$, indicating that resistance increases with temperature.

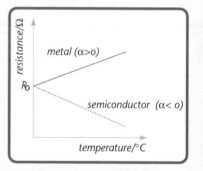

For semiconductors, as temperature increases electrons are liberated from their bound state and are free to conduct. This more than compensates for the increased vibrations of the atoms and therefore the current increases for a given potential difference, i.e. the resistance decreases. Therefore, for semiconductors $\alpha < 0$.

· A light dependent resistor (LDR) is a semiconductor component whose resistance changes with degree of illumination. As light intensity increases, resistance decreases.

Circuit symbol

6 Power (unit watt, W) is defined as **the rate of doing work, i.e. the rate at which energy is converted.**

$$P = Wd/t$$

Wd = work done in Joules (J) in time, t, in seconds

From the definition of the volt and the coulomb:

$$Wd = qV : q = It$$

$$\therefore \quad P = \frac{qV}{q/I} \quad \text{simplifying to give} \quad P = IV$$

substituting $V = IR \Rightarrow P = I^2R$

The amount of electrical energy transferred (Wd) in a given time (t) is given by the product of power and time, i.e.

$$Wd = IVt$$

7 The resistance of a wire is proportional to its length and inversely proportional to its cross-sectional area.

$$R \propto L \quad \text{and} \quad R \propto 1/A$$

combining gives: $R \propto \dfrac{L}{A}$ or $R = \dfrac{\rho L}{A}$

The constant of proportionality ρ is called resistivity and has the unit Ωm, where:

$$\rho = \frac{RA}{L}$$

Resistivity is a measure of a material's resistance to current and is independent of the dimensions of the sample. Resistivity can be defined as *the product of resistance and cross-sectional area per unit length.*

Have you improved?

1 a) Identify a component that could have its current and voltage characteristics illustrated by the *I–V* plot.

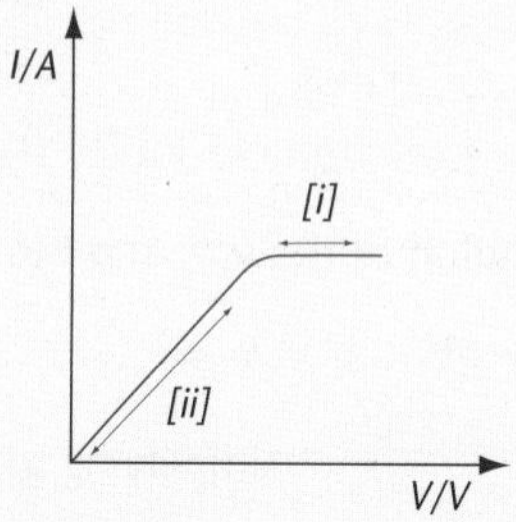

b) Select all the expressions from the list that would serve as appropriate labels for regions [i] and [ii]:

a) Maximum current
b) *I* decreases with *V*
c) Resistance is minimum
d) Resistance is constant
e) Resistance decreases with *V*
f) Current is constant

g) Minimum current
h) *I* increases with *V*
i) Resistance is maximum
j) Resistance increases with *V*
k) Resistance is zero
l) Potential difference is constant.

2 a) For the section of conductor shown, the number of electrons carrier per $m^3 = 5 \times 10^{29}$, the current is measured to be 2A, and the charge on each electron is 1.6×10^{-19}C.

Determine:

i) the amount of charge, in coulombs, contained in the conductor

ii) the drift velocity of the electrons.

b) If the PD across the sample is 2V, determine its

i) resistance

ii) resistivity.

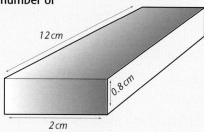

12 cm

0.8 cm

2 cm

$v = I/(nAe)$

$\rho = RA/L$

3 The graph shows the variation of resistance with temperature for a semiconductor (over a small temperature range). It is known that when a semiconductor is heated its atoms vibrate increasingly. This would seem to contradict the shape of the graph. Explain why the graph is correct.

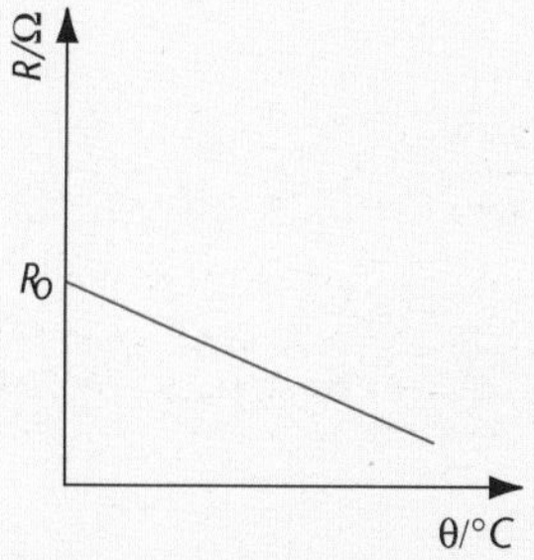

Electrical Circuits

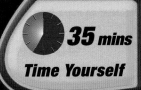

35 mins
Time Yourself

How much do you know?

1 Draw circuit symbols for a diode, variable resistor and a lamp.

2 Show how an ammeter and a voltmeter will be connected in a circuit to measure the current through and potential difference across a lamp connected to a DC supply.

3 Briefly describe the difference between EMF and PD.

4 Calculate R for the diagram opposite.

25 V
10 V
10 Ω
10 Ω
I
I
0.5 A
R

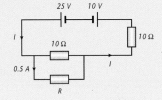

5 Calculate the total resistance of the arrangements shown:

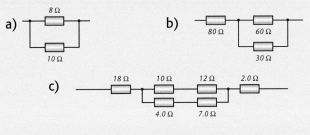

a)
8 Ω
10 Ω

b)
80 Ω
60 Ω
30 Ω

c)
18 Ω
10 Ω
12 Ω
2.0 Ω
4.0 Ω
7.0 Ω

6 Assume the cell in the circuit on the right has a negligible internal resistance. Calculate V_1 and V_2.

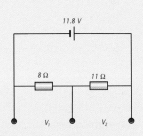

11.8 V
8 Ω
11 Ω
V1
V2

7 a) A cell has an EMF of 12 V and an internal resistance of 3 Ω. When placed in a circuit with a lamp, a current of 2 A is drawn from the cell. Calculate the output PD of the cell and the resistance of the lamp.

b) The lamp is replaced by a variable resistor, which can have values from 0 to 6W. Determine the maximum possible power that could be dissipated at the resistor.

8 An AC power supply produces the current variation as shown. Determine:

a) the peak and

b) RMS current.

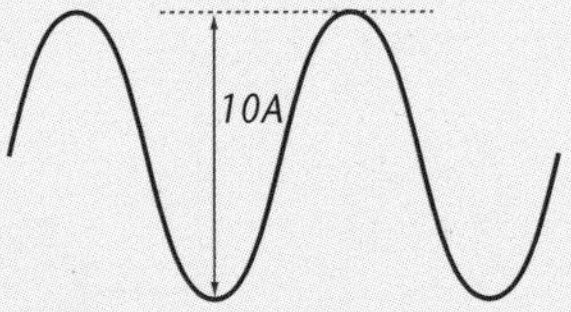

10A

Answers

1 see next page **2** see diagram **3** EMF is the energy supplied per unit charge in passing through a cell; PD is the electrical energy dissipated per unit charge passing between two points. **4.** 10Ω **5** a) 4Ω b) 100Ω c) 27Ω **6** V_1 = 5.0V V_2 = 6.8V **7** a) 6V, 3Ω b) 12W **8** a) 5A b) 3.5A

If you got them all right, skip to page 45

DAY 4

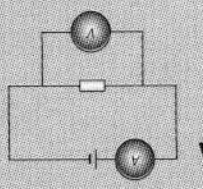

Spend no more than
45 mins
on this topic

Electrical Circuits

Learn the key facts

1 In circuit diagrams, components are given standard circuit symbols. Some of the more common symbols are presented below.

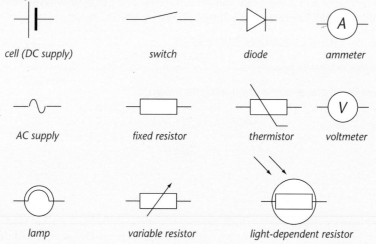

cell (DC supply)   switch   diode   ammeter

AC supply   fixed resistor   thermistor   voltmeter

lamp   variable resistor   light-dependent resistor

2 An **ammeter** is used to measure **currents** in electrical circuits. Ammeters are always **connected in series** with the component of concern. An ammeter has a very low resistance so that it does not add significantly to the resistance of the circuit.

A **voltmeter** is used to measure **potential differences** (voltages). Voltmeters are **connected in parallel** with the component across which the potential difference is to be measured. They have very high resistances to ensure that they draw minimal currents from the circuit.

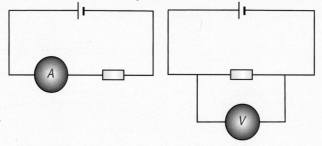

Electrical Circuits

Components connected in series all have the same current passing through them. Components connected in parallel all have the same potential difference across them.

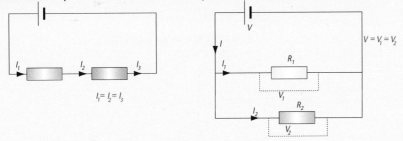

3 In the previous chapter, potential difference was defined as the electrical energy dissipated per unit charge flowing between two points.

Electromotive force (EMF) has the same defining equation.

e.g. $V = W/q$. EMF is related to PD but is nevertheless different.

EMF is defined as the energy supplied per unit charge in passing through a cell.

EMF should be considered to be the energy transfer occurring at the supply, i.e. charge gains electrical energy; potential difference is the transfer at the components, i.e. charge dissipates energy.

4 Kirchoff's Laws are used in DC circuit problems:

First Law: *The total current into a circuit junction is equal to the total current out of the junction.*

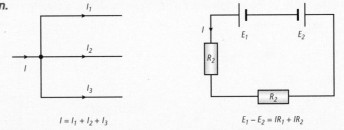

Second Law: *For any path in a circuit which forms a complete loop, the sum of the EMFs equals the sum of the products of current and resistance (allowing for polarity).*

5 Two resistors arranged in series will have a greater total resistance, R, than the same resistors arranged in parallel.

$$\text{series:} \quad R = R_1 + R_2$$

$$\text{parallel:} \quad \frac{1}{R} = \frac{1}{R_1} + \frac{1}{R_2}$$

DAY

4

Proof:

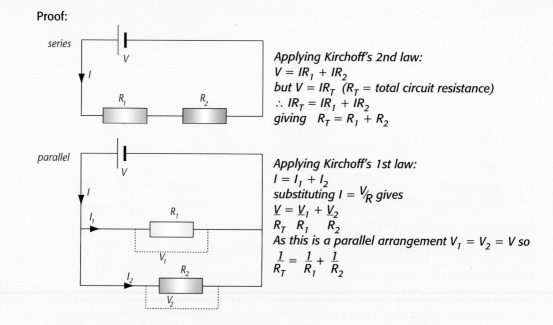

series

Applying Kirchoff's 2nd law:
$V = IR_1 + IR_2$
but $V = IR_T$ (R_T = total circuit resistance)
$\therefore IR_T = IR_1 + IR_2$
giving $R_T = R_1 + R_2$

parallel

Applying Kirchoff's 1st law:
$I = I_1 + I_2$
substituting $I = \frac{V}{R}$ gives
$$\frac{V}{R_T} = \frac{V_1}{R_1} + \frac{V_2}{R_2}$$
As this is a parallel arrangement $V_1 = V_2 = V$ so
$$\frac{1}{R_T} = \frac{1}{R_1} + \frac{1}{R_2}$$

6 Often voltage supplied by a cell is greater than the value required by a given component. A potential divider is used to divide the supply voltage into the values required.

The values of potential difference obtained at the output of a potential divider can range from zero to the value of the supply itself. The diagram shows a simple potential divider circuit.

For this circuit;

$V_{in} = V_1 + V_{out} = IR_1 + V_{out}$

$V_{out} = V_{in} - IR_1$

but $I = \dfrac{V_{in}}{R_1 + R_2}$ so $V_{out} = V_{in} - \dfrac{V_{in} R_1}{R_1 + R_2}$ which gives

$V_{\textbf{out}} = \dfrac{V_{in} R_2}{R_1 + R_2}$ and similarly $V_1 = \dfrac{V_{in} R_1}{R_1 + R_2}$ is obtained.

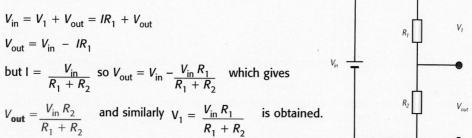

In general, for two resistors connected in series with a DC supply; *the potential difference across one resistor is equal to the product of its own resistance and the supply potential difference divided by the sum of the two resistances.*

For such an arrangement, the potential difference across one of the resistors will change if the value of the other resistor is altered. E.g. decreasing the resistance of R_1 will mean that a bigger share of the supply voltage is dropped across R_2, and vice versa. R_1 could be replaced by a component whose resistance is variable or dependent upon one of the physical conditions of the surroundings, e.g. a light-dependent resistor (LDR).

The resistance of R_1 would vary with light intensity; the greater the illumination the lower the resistance of R_1 and the greater the value of V_{out}. The system could be calibrated for known values of light intensity and the system then used as a light intensity meter. A similar arrangement where R_1 is replaced by a thermistor would allow temperature measurement in a similar manner.

7 If the potential difference across the terminals (terminal PD) of an isolated cell is measured, it has its maximum value and is called the EMF (electromotive force). It is this EMF (E) that drives current around a circuit.

When a current is drawn from the cell, the cell itself has a resistance; this is called the internal resistance of the cell. Some of the EMF has to overcome this resistance, resulting in a lower output PD for the circuit components.

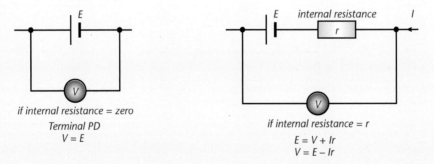

if internal resistance = zero
Terminal PD
$V = E$

if internal resistance = r
$E = V + Ir$
$V = E - Ir$

High-voltage (EHT) power supplies have a high internal resistance, which serves as a safety device. If the output is short-circuited, the high internal resistance will ensure that only a moderate current flows.

A car battery needs a low internal resistance as it needs to produce currents as high as 50 A in order to start a car engine.

For the circuit shown it is found that *the maximum power is dissipated by R, the load, i.e. delivered from the cell to the load, when R = r.* This is known as the Maximum Power Theorem.

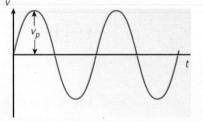

internal resistance, r

load resistance, R

8 The voltage of an AC power supply has a polarity that repetitively reverses in a sinusoidal manner.

When determining the effects of AC voltages, root mean square (RMS) values are often used:

$$V_{RMS} = \frac{V_p}{\sqrt{2}} \qquad I_{RMS} = \frac{I_p}{\sqrt{2}}$$

V_p = peak voltage

Electrical Circuits

Have you improved?

1 If the cell in the circuit below has zero internal resistance, calculate:
   a) The current leaving the cell.
   b) How much charge passes through the 8Ω resistor in 20 s.

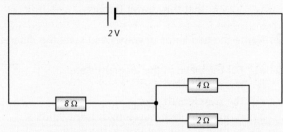

2 For the circuit on the right, calculate:
   a) The current in the 40 Ω resistor.
   b) The current in the 80 Ω resistor.
   c) The current in the 20 Ω resistor.
   d) The current leaving the cell.

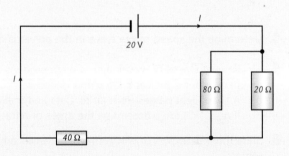

3 A cell with an EMF of 12 V and an internal resistance 2 Ω is connected across a lamp of resistance 46 Ω. Calculate the terminal PD when the lamp is glowing.

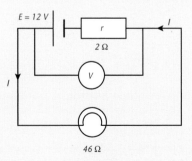

DAY

4

45

Basic Wave Properties

How much do you know?

1 Waves transfer _____ from one place to another. There is no net _____ of the _____ that they travel through.

2 Name the two kinds of waves and state the differences between them.

3 a)   For the wave in the diagram state:
   i)   the amplitude
   ii) the wavelength.
   b)   If one complete wavelength takes 0.005 s to pass a point determine the frequency of the wave.

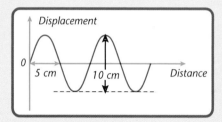

4 Determine the speed of the wave in the previous question.

5 a)   If water waves of wavelength λ were incident upon a gap of width d, state a condition needed for significant diffraction to occur.
   b)   A ray of light passes from air to glass and makes an angle of $50°$ with the interface. If $n_{glass} = 1.5n_{air}$, determine the angle of refraction.

6 State three examples of electromagnetic waves and give three properties that are common to each.

Answers

1 Energy/movement/medium **2** Longitudinal/transverse / the oscillations within a transverse wave are perpendicular to the direction of energy transfer, within longitudinal waves they are parallel
3 a) i) amplitude = 5 cm ii) λ = 10 cm b) 200 Hz **4** 20 ms^{-1} (λ = 10 cm) **5** a) d should be no bigger than λ. b) 25° **6** visible, infra-red, ultra-violet / speed, all transverse, travel in vacuum

If you got them all right, skip to page 52

Basic Wave Properties

Learn the key facts

1 Progressive (or travelling) waves transfer energy from once place to another, without transferring any of the medium (i.e. matter) that they travel through. The wave sets the particles of the medium to oscillate about fixed equilibrium positions along the path of the wave.

2 There are two classifications of waves:

Transverse waves are those that are created by vibrations that are perpendicular to the direction of energy transfer. The particles of the medium also oscillate perpendicularly to the energy transfer. Examples of transverse waves are electromagnetic waves, such as light.

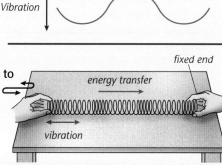

Longitudinal waves are a result of vibrations that are parallel to the direction of energy transfer. The particles in the waves' medium also oscillate parallel to the energy transfer. Examples are compression waves along a spring and sound waves.

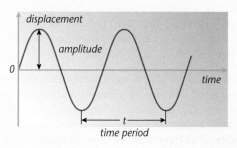

3 All travelling or progressive waves can be represented graphically from two perspectives:
(1) How it looks at an instant, i.e. how the displacement varies with position (a 'time snapshot').
(2) How it looks at a fixed position, i.e. how displacement varies with time (a 'space snapshot'). This type of wave is obtained on an oscilloscope (CRO), from which the time period and thus wave frequency can be calculated.

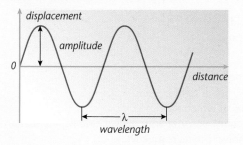

Fixed instant

Fixed position

Basic Wave Properties

An appreciation of the following wave terms is essential:

Displacement (s): Vector describing the position of a particle in the wave.

Amplitude (A): The maximum displacement of the particles in the wave medium from equilibrium. This is half the 'peak-to-peak' value.

Wavelength (l): The distance between corresponding adjacent points along the wave, i.e. the distance between crests.

Frequency (f): This is the number of wavelengths passing a point each second. Units: hertz (Hz).

Period (T): This is the time for a one wavelength to pass a point. Units: seconds (s). Related to frequency by T t 1/f.

Intensity (I): Power per unit area of the wave, proportional to the amplitude squared (A^2).

The Phase of a point in a wave is a measure of the fraction of a complete wave cycle completed from a chosen 'start' point. A complete oscillation is represented by 2p radians or 360°, half a wave by p or 180°, etc. Two different points are said to have the same phase if they are at the same stage in a wave cycle (e.g. both at maximum, or moving up through zero).

Wave crest is the maximum in amplitude and wave trough is the point of maximum negative amplitude.

Wavefront is a line joining points in a wave of equal phase – i.e. points 'in-phase' (e.g. all the wave crests).

A transverse wave can vibrate in any direction perpendicular to the wave direction. If the vibrations occur in only one plane in space, then the wave is called plane-polarised. Polarisation is a property of transverse waves alone, not longitudinal.

Visible light is polarised by passing it through a polaroid. The plane of polarisation can be rotated by liquid crystal displays (LCD) and sugar solutions, and this rotation is used in photoelastic stress analysis.

When longitudinal waves are represented graphically, remember the 'displacement' is *along* the direction of travel of the wave, not at right angles to it, as is the case for transverse waves. The points at which the pressure is highest (a compression) and lowest (a rarefaction) occur at zero displacements, when the neighbouring particles move in towards or out from these points respectively.

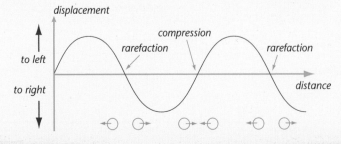

Basic Wave Properties

4 The speed of a wave is given by: $\boxed{v = f\lambda}$

Proof: speed = distance/time (for constant speed). In one time period ($T = 1/f$), a wave moves one wavelength (λ). Thus $v = \lambda/(1/f) = f\lambda$.

The factors that affect the speed of a wave are generally associated with the medium through which it passes, e.g.

waves on a string: $v = \sqrt{(T/\mu)}$
waves in a gas: $v \propto \sqrt{(P/\rho)}$ (constant T)

where T is tension (N); μ is the mass per length (kgm^{-1}); P the gas pressure (Pa) and ρ its density (kgm^{-3}).

5 Electromagnetic (EM) spectrum consist of electromagnetic waves, of which light is a small part.

They share the following properties:

- all are transverse waves
- all travel at the speed of light ($c = 3.0 \times 10^8 \text{ms}^{-1}$)
- all can be reflected, refracted and diffracted
- none require matter in order to propagate – they consist of oscillating electric and magnetic fields.

They only differ in their wavelengths, and frequencies (by $c = f\lambda$).

wavelength/m	Source	Use
10^{-15} — Gamma Rays	radioactive nuclei	medicine
10^{-12} — X-rays	fast electrons stopped rapidly	X-ray imaging
10^{-8} — Ultraviolet	Sun & very hot bodies	suntan/ kills germs/ photocells
10^{-7} — 400nm Blue Visible	Sun & very hot bodies	communication
10^{-6} — 800nm Red Infrared	hot bodies	heat detectors/ heating,
10^{-3} — Microwave	klystrons	cooking & communications
1 — Radio Waves	electric oscillations	radio/TV communications
10^{5} —		

6 When energy meets an interface, it may be absorbed, reflected and/or emitted. Light waves hitting a plane, shiny surface (e.g. a mirror) are reflected so that the angle of incidence, i, equals the angle of reflection, r.

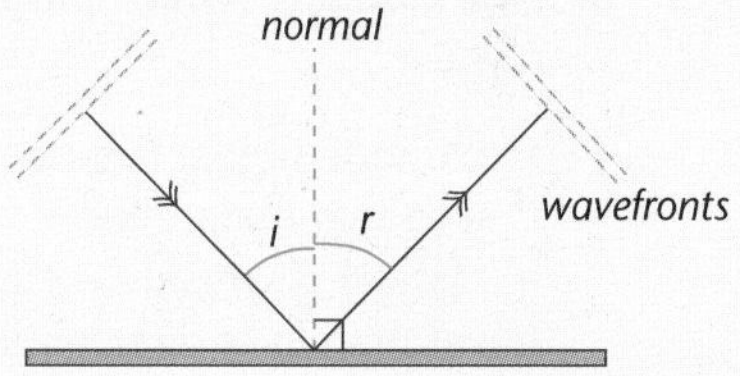

$\boxed{i = r}$

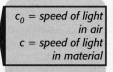

- In the diagram, the waves are represented by light rays, which point in the direction of motion (and are at 90° to the wavefronts).
- N.B. ALL angles are measured from the 'normal' (i.e. a line at 90° to the surface) in optics.

Refraction occurs when a wave travels between media of differing densities, as it changes speed and may change direction. The frequency, however, remains unchanged.

c_0 = speed of light in air
c = speed of light in material

For light, materials are characterised by their refractive index (n), which is a measure of the speed of light in the material (c) – a bigger n implies a slower wave:

$n \text{ (air)} = 1.0$
$n \text{ (glass)} = 1.5$
$n \text{ (water)} = 1.33$

$\boxed{n = c_0/c}$

Basic Wave Properties

When light passes from a less dense to a more dense medium it slows down and diverts towards the normal (i.e. the line at 90° to the interface). The amount that the light decreases its speed depends upon the refractive index of the two materials it passes between, as given by Snell's Law:

$$\frac{\sin\theta_1}{\sin\theta_2} = \frac{c_1}{c_2} = \frac{n_2}{n_1}$$

where n_1 = refractive index of medium 1; n_2 = refractive index of medium 2; c_1 = speed of light in medium 1; c_2 = speed of light in medium 2.

N.B. Sound travels faster in denser materials; water waves travel faster in deeper water; and paraffin wax has a high n for microwaves.

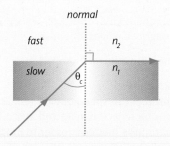

- Total internal reflection (TIR) occurs when the incident angle exceeds the 'critical angle'. This critical angle, θ_c, occurs when the reflected ray just manages to emerge along the interface between the two media.

$$\sin\theta_c = n_2/n_1 \qquad \text{for} \qquad n_2 < n_1$$

TIR is used in optical fibres to send communications via light rays down glass fibres. Step-index fibres have an abrupt change in refractive index between the core (with a larger refractive index) and the cladding (which protects the core and provides the change in n values). These can be of two types:

1 Multimode fibres: large cores ($d \sim 50\,\mu$m). The time period of an input 'pulse' is increased due to two main effects:
   a) Modal dispersion – Light that is reflected by TIR takes much longer to travel than light directed along the core axis. This can be overcome using a different type of fibre, a graded-index multimode fibre, where the change in the value of n occurs gradually.
   b) Material dispersion – shorter wavelengths of light travel more slowly than longer wavelengths.
2 Monomode fibres: small cores ($d \sim 5\,\mu$m). A laser containing only a few wavelengths (thus decreases material dispersion) is directed along the core axis, reducing reflections (and thus modal dispersion).

- Diffraction occurs if a wave moves past an obstacle or through a small gap (of a size comparable to λ) when it spreads out into the shadow region of the slit, e.g. water waves, sound waves, microwaves.

Basic Wave Properties

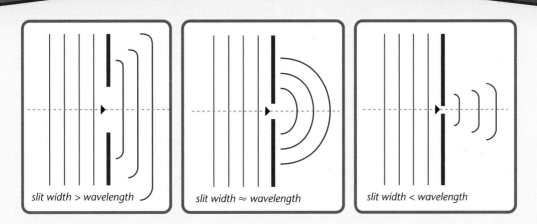

slit width > wavelength   slit width ≈ wavelength   slit width < wavelength

The narrower the gap and larger the wavelength, the greater the diffraction.

When light falls on a narrow slit, a diffraction pattern is observed. This consists of a series of bright and dark fringes; the intensity (brightness) of the central fringe is the greatest. The dark fringes occur when:

$$\sin\theta = \lambda/a$$

where θ = angle of diffraction, λ = wavelength of light and a = slit width.

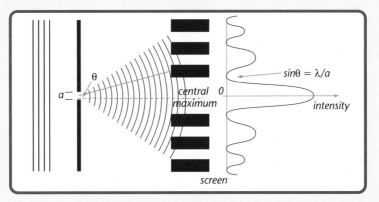

When loudspeakers and satellite dishes for emitting signals are designed, their widths are chosen so that a lot of diffraction occurs (i.e. $\lambda \approx a$). Likewise, long wavelength radio signals will reach the valley floor in mountainous regions, as they diffract around the edges of the hills.

Basic Wave Properties

Have you improved?

1 a) Explain the term frequency as applied to wave motion.
   b) What is polarisation? Why can longitudinal waves not be polarised? Give an example of a wave that can be polarised.

2 If a wave on a plucked string has a speed v, determine the new speed, in terms of v, if the string is replaced by one of the same length, twice the mass and half the tension.

$v = \sqrt{(T/\mu)}$

3 a) List three properties common to all EM waves. Name one way in which they differ.
   b) Identify from which region the following wavelengths of EM radiation originate: 0.1 pm; 1 km; 0.1 nm; 50 nm; 1 cm; 300 nm; 0.1 mm.

4 Determine the magnitude of the gaps that would cause significant diffraction for:
   a) a sound wave travelling at $330 \, ms^{-1}$ with a frequency of 500 Hz
   b) a radiowave of speed $3 \times 10^8 \, ms^{-1}$ and with a frequency of 1×10^7 Hz
   c) a visible light wave with frequency of 6×10^{14} Hz.

diffraction is significant when $a \simeq \lambda$

Use $c = f\lambda$

5 A light ray travelling through air enters a rectangular glass block at an angle of incidence of 40°.
   a) Determine the angle of refraction in the glass block. (Speed of light in air $3 \times 10^8 \, ms^{-1}$; refractive index of glass = 1.5)
   b) The refracted ray then strikes the second face of the glass block. What is its angle of incidence?
   c) What is the critical angle for the glass–air interface?
   d) Does the ray emerge from the glass block at this interface? Explain.

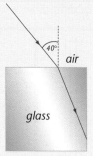

6 a) Calculate the critical angle for a step index fibre with a core refractive index of 1.472 and cladding refractive index of 1.455.
   b) Sketch a fibre which is in a straight line and show the paths of both the fastest and slowest routes light can take.

Further Waves

How much do you know?

1 When two waves meet at a point, the total displacement is given by _____ the _____ of each wave at the point. This is called _____.

2 a) Interference is the _____ of two or more waves, which have the same _____ and are coherent.

b) P and Q generate waves that are in phase. If the waves have a wavelength of 0.25 m determine the nature of the interference at X.

3 Consider the Young's double-slit experiment shown in the diagram. Determine the separation of the observed fringes.

4 Light of wavelength 800 nm is incident at 90^0 to a diffraction grating with 100 lines per millimetre. What is the angle through which the second-order beam is deflected?

5 a) Label the diagram with nodes and anti-nodes.

b) If the wave speed = $250\,ms^{-1}$, calculate the frequency for the harmonic shown in the diagram.

c) Determine the fundamental frequency.

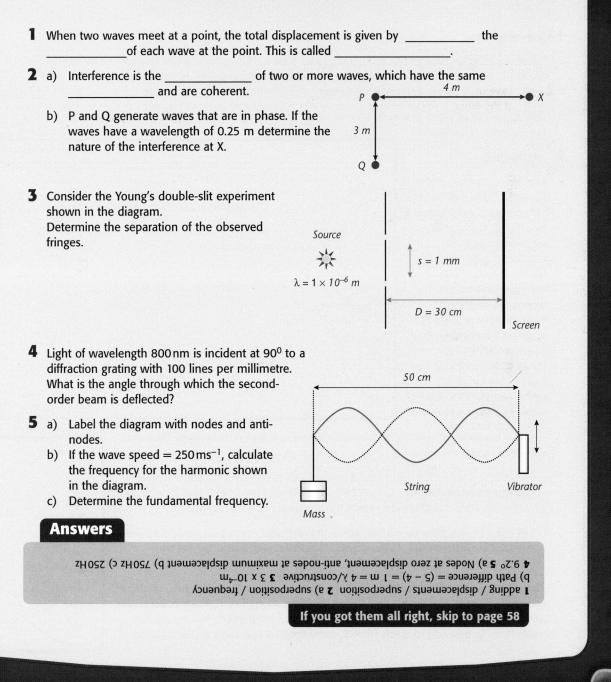

Answers

1 adding / displacements / superposition **2** a) superposition / frequency
b) Path difference = $(5 - 4) = 1\ m = 4\ \lambda$/constructive **3** 5×10^{-4}m
4 9.2^0 **5** a) Nodes at zero displacement, anti-nodes at maximum displacement b) 750 Hz c) 250 Hz

If you got them all right, skip to page 58

Further Waves

Learn the key facts

1 Superposition

If two waves pass through the same point, they pass on through the point unaffected. While they are at the point they combine by superposition: the resultant displacement equals the vector sum of the individual displacements at that point.

This applies to all types of waves, whether they are transverse or longitudinal, e.g. water waves, microwaves, sound waves.

2 Interference

In practice, waves consist of many crests and troughs, rather than a single wave. If two 'wave trains' of the *same frequency* and constant phase difference (i.e. coherent) overlap, then interference occurs, where the resultant displacement is given by superposition of the two 'wave trains'. This can occur for all types of waves, e.g. water waves in ripple tanks, sound or EM waves (such as light, microwaves).

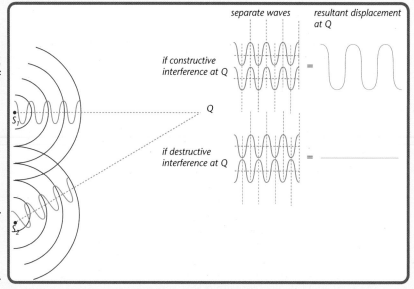

E.g. Two separate sources, S_1 and S_2, emit spherical wavefronts of equal frequency and amplitude.

* At certain points they reinforce (i.e. arrive in 'phase' – crest on crest) to give constructive interference. These points form 'lines' (see lines A and C) where the resulting wave oscillates with twice the amplitude of either source.

* At other points, they cancel (i.e. arrive 'out of phase' – crest on trough) to give destructive interference. These points form 'lines' (see line B) where the resulting wave has zero amplitude.

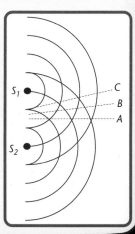

Along line A both waves have travelled the same number of wavelengths, so there is no difference in their phase or distance travelled. Along line C, every point is one whole wavelength closer to S_1 than S_2. Constructive interference still occurs, but crests from S_2 that add to crests from S_1 were emitted one cycle earlier than that from S_1. We say there is a phase difference of 1 oscillation (i.e. 360° or 2π radians). 'In phase' simply means that the phase difference is a whole number of oscillations. Conversely, 'out of phase' means a phase difference of a whole number of half oscillations (e.g. 180°, 540°).

In the particular case where two waves start in phase, rather than considering their phase differences to determine how they interfere, we consider the difference in distance they travel to the interference point, called the path difference:

$$PD = S_1Q - S_2Q$$

$PD = n\lambda \quad \Rightarrow \quad$ constructive interference (e.g. 0, λ, 2λ, 3λ, ...)

$PD = (n + \frac{1}{2})\lambda \Rightarrow$ destructive interference (e.g. $\lambda/2$, $3\lambda/2$, $5\lambda/2$, ...)

Where n is an integer 0, 1, 2, 3, ...

If the interference pattern from two sources is viewed along a line at 90° to line A (in the previous diagram), a series of 'fringes' is observed where constructive interference occurs. For water waves these appear as regions of large disturbance interspaced with calm regions; while for light there are successive patches of light and dark.

3 Young's double-slit experiment is used to obtain the interference pattern for light.

sodium lamp
single slit
double slit
beams interfere in this region
screen
diffracted beam from single slit
interference pattern
overlapping diffracted beams from double slits (coherent sources)
dark fringes
bright fringes

- This experiment is important as it demonstrates that light can behave as a wave – as all other waves show interference (see 'Quantum Physics' for evidence of light acting as a particle).
- To produce a stationary interference pattern the wave sources must be:
  Coherent (i.e. have a constant phase difference, usually chosen to be zero – sources are 'in phase').
  Young ensured this by putting a single slit between the light source and double slits, or alternatively a laser is used.
  Monochromatic (i.e. single constant wavelength or frequency). Otherwise, the interference pattern becomes 'smeared out' due to different wavelengths having different fringe separations.
- If the slits are of comparable size to the wavelength of light (10^{-7} m), the light diffracts through each slit and will overlap. In this overlap region, interference will occur.
- Alternate bright and dark 'fringes' of equal spacing are seen on the screen, corresponding to constructive and destructive interference respectively. If one slit is covered up, the interference pattern disappears.
- The distance between the centres of adjacent bright, or dark, fringes is called the fringe spacing, x;

$\lambda \approx xs/D$

where s (1 mm) is the slit separation, D (1 m) the perpendicular distance of the screen from the slits, and λ (10^{-7} m) the wavelength of the light.

- The fringe spacing equation indicates that longer wavelengths have larger fringe spacings. Thus, if white light is used, multicoloured fringes are seen, with blue (shorter wavelength) having smaller fringe spacings than red.
- To measure the fringe separation, measure the separation of several fringes (e.g. 10) with a travelling microscope and then calculate the separation of one fringe.
- As the light is diffracted by the single slit, the fringe pattern is modified by the single-slit diffraction pattern as shown opposite.

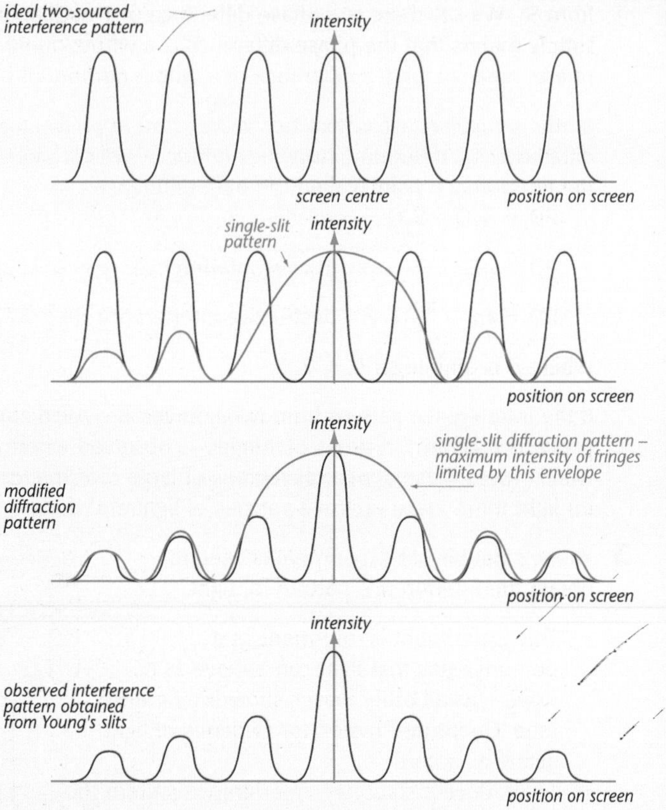

Although these details are specific to the interference of light, the principles apply to all two-source interference.

4 Diffraction gratings can contain any number of parallel, equally sized and spaced slits (e.g. two slits form a the simplest diffraction grating). For an incident parallel beam at 90° to the grating, the emergent beams are only seen in directions where:

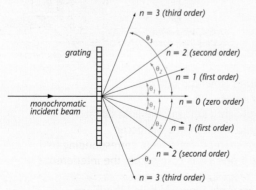

$$d\sin\theta = n\lambda$$

where d is the spacing of the slits, n the order of the diffracted beam (0, 1, 2, ...), λ the wavelength of the incident light and θ the angle the beam has been diffracted through.

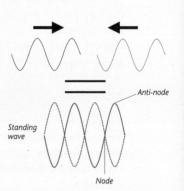

Commonly, this is used to determine the wavelength of the incident light (see 'Quantum Physics').

5 Stationary (standing) waves are produced when two progressive waves travelling in different directions of the same frequency, speed, wavelength and similar amplitude, superimpose. Both waves must be transverse or both must be longitudinal.

Unlike progressive waves, where each point along the wave experiences the full amplitude of the oscillation during a cycle, points in standing waves have different amplitudes. At fixed positions there will be regions of constructive interference, anti-nodes, and regions of destructive interference, nodes (no displacement). The distance between two successive nodes (or anti-nodes) is half a wavelength, $\lambda/2$. All points between two successive nodes are in phase (i.e. are at the same stage in their cycle of oscillation).

Progressive **wave** (travelling)	Standing **wave** (stationary)
transports energy	stores energy
all points have same amplitude	position – varying amplitude
phase difference varies $0 \rightarrow 360°$	phase difference only 0 or $180°$

Stationary waves on strings are formed when a wave travels along a string that is under tension and fixed at both ends: the reflected wave will combine with the incident wave to form a standing wave. Only certain certain frequencies resonate to give a standing wave with a large amplitude.

- The lowest frequency at which this occurs is the fundamental frequency (f_0).
- Standing waves may also be formed at whole number multiples of f_0 and are called harmonics or overtones. The first overtone has twice the frequency and half the wavelength of the fundamental.

overtone	string	closed pipe	open pipe
fundamental	$\lambda_2 = 2l$	$\lambda_1 = 4l$	$\lambda_2 = 2l$
1st overtone	$\lambda_2 = l$	impossible	$\lambda_2 = l$
2nd overtone	$\lambda_3 = 2l/3$	$\lambda_3 = 4l/3$	$\lambda_3 = 2l/3$
3rd overtone	$\lambda_4 = l/2$	impossible	$\lambda_4 = l/2$

- There is always a node at the fixed ends of the string.
- This is an example of a transverse standing wave.
- These are used in musical string instruments. Different pitches (frequencies, i.e. 'notes') are created by using strings with different wave speeds (by varying the string tension and/or mass per length).

Standing waves can also be created in columns of air in pipes (e.g. wind instruments, organ pipes). Closed ends have nodes and open ends have anti-nodes (see table).

Further Waves

Have you improved?

1 If the two waves are incident at the same point at the same time, sketch the result of the superposition that occurs.

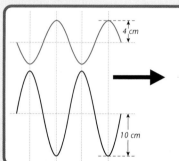

2 If A and B are the source of waves that are in phase, with $\lambda = 2\,m$, determine the nature of the interference at:
a)   Point X.
b)   Point Y.
c)   Point Z.

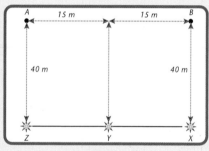

DAY

5

3

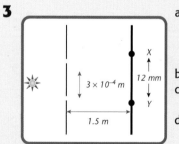

a)   For the Young's double-slit experiment shown in the diagram, 30 fringes are observed between X and Y. Determine the wavelength of the light emitted by the source.
b)   How could the fringe spacing be increased?
c)   What would be the effect of increasing the slit widths, while keeping the slit separation the same?
d)   What would be observed if white light were used rather than monochromatic light?

4 a)   For the string shown in the diagram, determine the wavelength and frequency of the three lowest-frequency standing wave modes. (Wave speed on string = $250\,ms^{-1}$)
b)   What are the wavelength and the three lowest frequency standing wave modes in an open pipe of the same length, 50 cm?

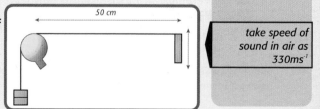

take speed of sound in air as $330\,ms^{-1}$

Heat

20 mins
Time Yourself

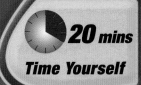

How much do you know?

1 A 2 kg sample of aluminium is warmed from 0 °C to its melting point of 660 °C. Heating continues until all the aluminium is molten. Determine the total energy required, assuming no heat losses, for this process.

For aluminium: $l = 412000\,Jkg^{-1}$ $c = 100\,Jkg^{-1}$

2 For a 96 g sample of oxygen molecules with molar mass 32 g, determine:
($N_A = 6.0 \times 10^{23} mol^{-1}$)
a) The number of moles of oxygen molecules.
b) The number of molecules.

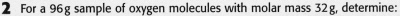

3 For the molecules in the diagram determine:
a) The average velocity.
b) The mean square speed.
c) The root mean square speed.

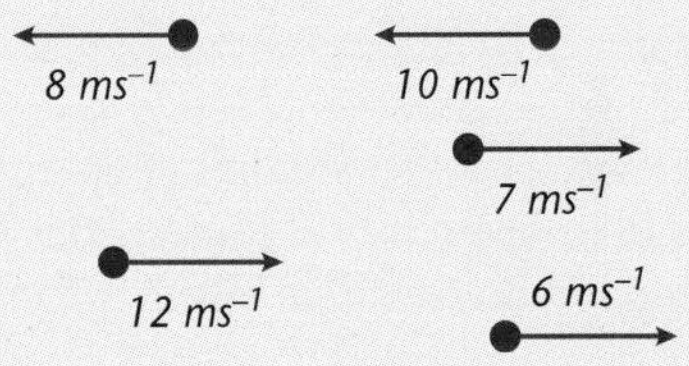

8 ms⁻¹
10 ms⁻¹
7 ms⁻¹
12 ms⁻¹
6 ms⁻¹

4 Boyle's Law states that for a fixed ________ of gas at a constant ________ the product of ________ and ________ is ________.

5 A vessel of volume $30 \times 10^{-2} m^3$ contains gas at a pressure of $2000\,Nm^{-2}$ at a temperature of 330 K. Calculate:
($R = 8.31\,JK^{-1}mol^{-1}$; $N_A = 6.0 \times 10^{23} mol^{-1}$)
a) The number of moles of gas.
b) The number of molecules of gas.

6 A 2 kg sample of gas contains molecules with a mean square speed of $2500\,ms^{-1}$. If the container has a volume of $2 \times 10^{-3} m^3$ determine the pressure of the gas.

7 Write down the freezing and boiling points of pure water in Kelvin.

Answers

1 $9.56 \times 10^5 J$ **2** a) 3 moles b) 9.1×10^{23} molecules **3** a) $1.4\,ms^{-1}$ to the right b) $78.6\,m^2s^{-2}$ c) $8.9\,ms^{-1}$ **4** Mass/temperature/pressure/volume/constant **5** a) 0.22 moles b) 1.3×10^{23} molecules **6** $8.33 \times 10^5 Nm^{-2}$ **7** 273 K, 373 K

If you got them all right, skip to page 63

Heat

Learn the key facts

1 Specific heat capacity (c) is the property of a substance that allows two bodies of the same mass to receive the same amount of heat energy but increase their temperatures by different amounts. Specific heat capacity is defined as *the heat (energy in joules) required to increase the temperature of 1kg of a substance by 1°.*

$$E = mc\Delta\theta$$

Heat capacity (C) is the total heat energy required for a whole body (rather than 1kg) to increase its temperature by 1°.

$$E = C\Delta\theta$$

Latent heat is the amount of energy that has to be given to a body to change its state (solid to liquid, liquid to gas). This is independent of the energy required to raise its temperature to that at which it changes state, or any subsequent increase.

Specific latent heat (l) is the amount of heat energy required to cause 1kg of a substance to change state.

$$E = ml$$

Specific latent heat of fusion is the amount of energy required to change 1 kg of a solid at its melting point to a liquid.

Specific latent heat of vaporisation is the amount of energy required to change one kilogram of a solid at its boiling point to a gas.

2 An appreciation of the following terms is essential when considering gases.

- The mole: **The amount of a substance that contains Avogadro's number of particles.**
- Avogadro's number (N_a) = 6.022×10^{23} mol^{-1}: **This is the number of particles in one mole of a substance.**
- Molar mass: **This is the mass of one mole (6.022×10^{23}) of atoms of a substance.**

3 For *n* molecules/atoms of a gas with speeds $c_1, c_2, c_3, \ldots c_n$

mean speed $= c = \dfrac{c_1 + c_2 + c_3 + \ldots c_n}{n}$

mean square speed $= \overline{c^2} = \dfrac{c_1^2 + c_2^2 + c_3^2 + \ldots c_n^2}{n}$

root mean square speed $= c^2 = \dfrac{c_1^2 + c_2^2 + c_3^2 + \ldots c_n^2}{n}$

> E = heat energy (J)
> m = mass (kg)
> $\Delta\theta$ = temperature change (°C or K)
> c = specific heat capacity $Jkg^{-1}K^{-1}$

> E is the heat energy required to change the state of a substance of mass, m, and specific latent heat, l.

4 The three laws that describe the behaviour of gases are:

Boyle's Law: $P \propto \dfrac{1}{V}$ fixed mass
constant temperature

Charles' Law: $V \propto T$ fixed mass
constant pressure

Pressure Law: $P \propto T$ fixed mass
constant volume

> P = pressure (Nm^{-2})
> V = volume (m^3)
> T = temperature (K)

5 a) An ideal gas is one that obeys each of the gas laws. No gas is completely ideal, but as a theoretical concept, it is useful when modelling the behaviour of gases.

The following points are assumed true for an ideal gas:

- There are a large number of molecules moving randomly.
- The molecules continuously collide with each other and the walls of the container.
- The volume of the molecules is negligible compared to that of the container.
- Apart from during collision the forces between molecules are negligible.
- The contact time during a collision is negligible compared with the time spent between collisions.
- All collisions (on average) are elastic (KE conserved).

The equation which describes the behaviour of an ideal gas, is called the equation of state of an ideal gas.

> $PV = nRT$
> or
> $PV = NkT$

For a particular gas this leads to:

$$\frac{P_1 V_1}{T_1} = \frac{P_2 V_2}{T_2}$$

The pressure of an ideal gas is related to the mean square speed of its molecules and its density by:

$$P = \frac{1}{3} \rho \overline{c^2}$$

The average Kinetic energy per molecule of and ideal gas is given by:

$$KE_{molecule} = \frac{1}{2} m \overline{c^2} = \frac{3}{2} kT$$

per mole:

$$KE_{mole} = \frac{1}{2} M \overline{c^2} = \frac{3}{2} RT$$

where m = mass of one molecule, M = mass of one mole

b) The (Kelvin) absolute scale of temperature is a theoretical temperature scale, which has its zero point as the lowest temperature that can be reached in theory (never achieved).

Another reference point on this scale is the triple point of water. This is the unique temperature at which water exists in the liquid, gas, and solid phases simultaneously.

One unit on this scale is one Kelvin (1 K). The lowest temperature (absolute zero) = 0 K and the triple point of water = 273.16 K (273 K used for most applications).

A consequence of defining these reference points is that 1 K = 1°C.

$$T_k = T_c + 273$$

n = number of moles (mol)
N = number of molecules
R = universal molar gas constant (8.31 JK⁻¹mol⁻¹)
k = Boltzman's constant (1.38 × 10⁻²³ JK⁻¹)

ρ = gas density (kgm⁻³)

Have you improved?

1 A 1 kg block of ice at 0 °C has 2×10^6 J of energy supplied to it.
Calculate:
a) The energy required to melt all of the ice.
b) The energy required to raise the water to its boiling point.
c) How much of the total energy supplied remains after processes a) and b).
d) What mass of steam is produced by this remaining energy.

2 a) Identify each symbol in the equation $PV = nRT$.
b) A gas container of volume 1 m^3 contains a gas with a pressure of
1.5×10^7 Nm^{-2}. If the molar mass of the gas is 32 g and its temperature =
27 °C determine:
i)   The number of moles of gas in the container.
ii)   The mass of gas in the container.
iii) If the container has a safety valve that releases gas when the internal
pressure exceeds 30×10^5 Nm^{-2} determine the mass of gas released
if the temperature of the container rises to 54 °C.

3 State three conditions (other than obeying the gas laws) that an ideal gas has
to satisfy.

4 An ideal gas at a pressure of 2×10^5 Nm^{-2} is trapped within a thermally
isolated container of volume 5×10^{-2} m^3 and contains 30 g of gas at 300 K.
Determine:
a) The number of moles of gas present.
b) The molar mass of the gas.
c) The mean square speed of the molecules.

5 A gas is held in a vessel of volume 0.01 m^3 under a pressure of 5×10^6 Nm^{-2}
at 350 °C. If the gas is released so that it is at atmospheric pressure
(1×10^5 Nm^{-2}) and 300 °C then determine its new volume.

For water: specific latent heat of fusion = 33.44 × 10⁴ Jkg⁻¹ specific latent heat of vaporisation = 226.1 × 10⁴ Jkg⁻¹ specific heat capacity = 4200 Jkg⁻¹

R = 8.31 JK⁻¹mol⁻¹

use n = pV/RT

1

2

3

4

DAY

5

6

7

Solids

How much do you know?

1 If a _____ is applied to a sample causing its length to decrease, then the sample is under _____. If the sample had extended it would have been under _____.

2 The stress–strain curve was obtained by increasing the tension of a wire until it snapped. Use the graph to estimate the total work done in snapping the wire. The wire had an initial length of 5 m and a radius of cross section of 0.5 mm.

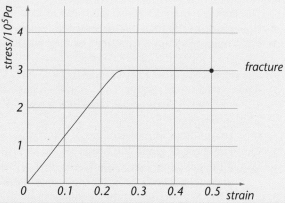

3 A metal rod of length 10 m has a radius of cross-section of 5 mm. A force of 10^5 N is applied along its length causing an extension of 0.1 mm. Calculate the rod's stress, strain, and Young modulus.

4 Which of the materials, a, b, or c, are:
i) toughest ii) stiffest iii) strongest iv) most brittle v) most ductile?

5 Sketch (in 2D) an example of the atom arrangement in an amorphous and a crystalline solid.

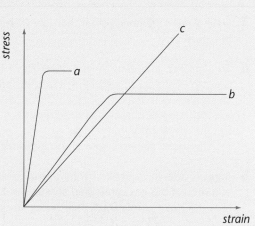

Answers

amorphous   *crystalline*

iv) c v) b **5** see diagrams
strain = 1 × 10^{-5}/Young modulus = 1.2 × 10^{14}Nm^{-2} **4** i) b ii) a iii) c
1 force/compression/tension **2** 0.44J **3** stress = 1.3 × 10^9Nm^{-2}/

If you got them all right, skip to page 70

64

Solids

Spend no more than
45 mins
on this topic

Learn the key facts

1 When a force is applied to a sample of material it will deform, i.e. change its shape and/or size.

If the applied force(s) reduces the length of the sample then the sample is under compression. An increase in the length means that the sample is under tension.

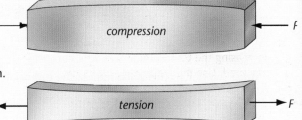

2 Elastic deformation occurs when a material returns to its original shape and size on removal of deforming forces. No bonds between atoms are broken.

Plastic deformation occurs when a permanent extension of the material remains following the removal of the extending force. Small (sometimes zero) increases in force can cause a large extension. Bonds between atoms are broken and reform between different atoms.

Hooke's Law is obeyed when the extension of a material is proportional to the resultant deforming force, and is defined by the equation:

$$F = kx$$

F = deforming force (N), x = extension (m), k = stiffness constant (Nm^{-1})

A plot of force against extension is often useful when comparing the properties of materials.

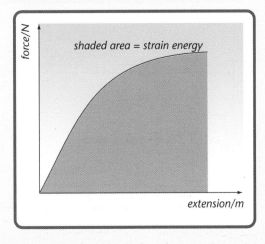

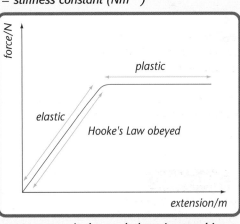

Strain energy is the work done in stretching a sample (stored as potential energy in the sample) and is given by the area under the corresponding force–extension graph.

If Hooke's Law is obeyed:

$$E = \frac{1}{2} Fx$$

DAY

5

3 When considering force–extension it is difficult to distinguish between properties that are a result of the material itself and those due to its dimensions. The following terms are related to the type of material being investigated rather than its shape or size.

Stress (units Nm^{-2} or Pa): This is the ratio of the force applied to a sample and the original (before stretching) cross-sectional area of the sample.

$$\sigma = \frac{F}{A}$$

Strain (no units): This is the ratio of the extension and the original length of the sample.

$$\varepsilon = \frac{x}{L}$$

Young modulus, E (units Nm^{-2}): This is the ratio of stress to strain.

$$E = \frac{\sigma}{\varepsilon} \quad \text{this leads to} \quad E = \frac{Fl}{xA}$$

The value of the Young modulus is determined from the gradient of the portion of a stress–strain plot where stress α strain (Hooke's Law obeyed) and is a measure of the stiffness (resistance to bending or stretching) of the material.

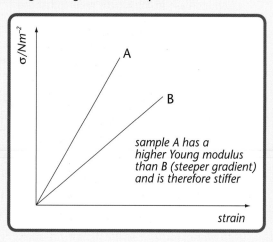

sample A has a higher Young modulus than B (steeper gradient) and is therefore stiffer

4 Many properties can be determined from a stress–strain graph.

A: Limit of proportionality. The maximum applied stress within which stress α strain i.e. Hooke's Law obeyed.
B: Elastic limit. The material retains its elastic behaviour for stresses up to this value.
C: Yield point (yield stress). The minimum stress at which the material exhibits plastic behaviour, i.e. very small (or zero) increases in stress produce large, and permanent, increases in strain.
D: Ultimate tensile stress. This is the greatest stress that a material can withstand before fracture (breaking).

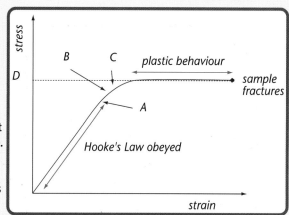

The following terms are commonly used when describing materials.

Strong: A strong material can withstand a large stress before fracture, i.e. has a large ultimate tensile stress. E.g. steel.

Weak: Opposite of strong. E.g. balsa wood.

Stiff: A stiff material offers a large resistance to changes in its shape or size (stretching or bending). Large stresses are required to produce small strains. E.g. high-carbon steel, diamond.

Flexible: Opposite of stiff. E.g. polythene.

Tough: Tough materials undergo a significant period of plastic deformation before fracture. A tough material requires a large amount of energy per unit volume to break it – i.e. it would have a large area under its stress–strain graph.

Brittle: Brittle materials demonstrate little or no plastic deformation prior to fracture. Brittle materials fracture due to the rapid growth of cracks and are generally stronger in compression than they are under tension. E.g. glass.

Ductile: Ductile material can be drawn out into long wires. For a material to be classed as ductile it needs to demonstrate a large degree of plastic behaviour before fracture, i.e. permanent extensions > 20% of original length. Ductile materials are also tough. E.g. copper.

Malleable: Malleable materials can be hammered into shape and beaten into very thin sheets. E.g. gold.

The area under a stress–strain graph gives the energy per unit volume stored in a material under tension.

For certain materials, e.g. rubber, the stress for a given strain during loading (increasing force) is greater than that during unloading. This is seen on a stress–strain graph as a hysteresis loop.

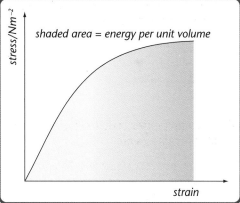

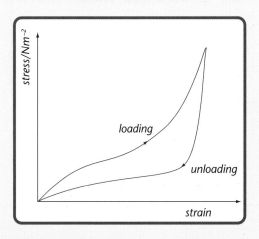

The area contained within the loop gives the energy, per unit volume of sample, released by the sample during one cycle of loading and unloading. Rubber can be felt to get warm as it is repetitively stretched and released.

Typical stress–strain graphs

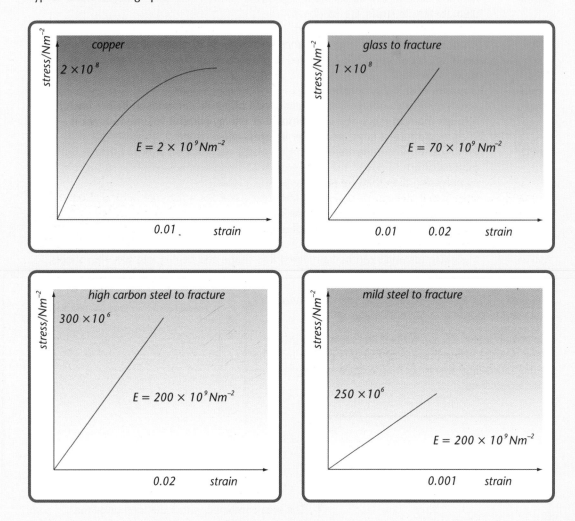

5 A material with a crystalline structure has atoms arranged in a regular pattern. A unit cell, e.g. a cube, is repeated throughout the material; there is long-range order.

metal crystal

graphite

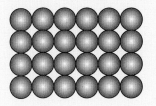

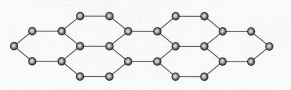

A polycrystalline material consists of tiny grains (less than 1 mm across). Within these grains, atoms are arranged in a regular crystalline manner; the grains however have no order with respect to each other. The grains are randomly orientated.

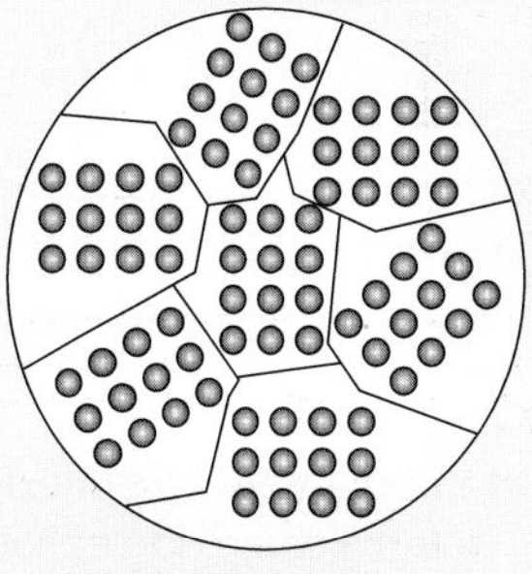

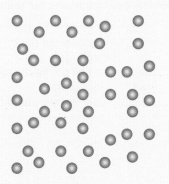

The atoms of an amorphous solid are disordered (randomly arranged). Although these materials are solids they have an atomic arrangement very similar to that of liquids and demonstrate slight liquid properties, e.g. they flow (very slowly). E.g. natural rubber, glass, ceramics.

amorphous

Have you improved?

1 The following results were obtained for an experiment where increasing force was applied to a metal wire.

force/N	extension/mm
10	0.5
20	1.0
30	1.5
40	2.0
50	2.5
60	8.0
70	20.0
80	50.0

Plot the results on a force–extension graph.
Mark clearly on your graph the region where the sample exhibits plastic deformation.

2 A metal bar can withstand maximum stress and strain of $150 \times 10^6\,\mathrm{Nm^{-2}}$ and 0.002 respectively before fracture. The bar is of rectangular cross section with sides 3 mm and 8 mm and has an unextended length of 3 m. Calculate:
i)   the maximum force that the bar can withstand
ii)   the maximum extension of the bar
iii) the Young modulus of the material assuming
      Hooke's law is obeyed until fracture
iv)   the energy stored in the sample just before breaking.

$F = \sigma A$
$x = \varepsilon L$
$E = \sigma/\varepsilon$
$Wd = \frac{1}{2}Fx$

3 Which of the expressions i) to v) could be associated with the solid structure types A, B and C.

A: Crystalline
B: Polycrystalline
C: Amorphous

i)   long-range order
ii)   no order
iii)   small grains of ordered atoms
iv)   e.g. glass
v)   e.g. sodium chloride

4 Sketch, on a single set of axes, stress–strain graphs for the following types of material:
a) brittle
b) tough
c) strong
d) flexible.

DAY

5

Quantum Physics

How much do you know?

1 When light with a _____ above the threshold value falls upon a metal surface then _____ will be emitted. The existence of this threshold value, along with the electron _____ energy dependence upon frequency, led to the _____ model of electromagnetic radiation.

2 a) $KE = hf - \phi$
   Explain the meaning of the terms in this equation.
   b) Photons of frequency $2 \times 10^{15}\,Hz$ fall upon a metallic surface of work function 1 eV. Determine the maximum kinetic energy of any ejected electrons ($h = 6.63 \times 10^{-34}\,Js$).
   c) Determine the threshold frequency for a metal surface of work function 8eV.

3 Give one phenomenon that supports a wave model for electromagnetic radiation and one for a particle model.

4 An electron, of mass $9.1 \times 10^{-31}\,kg$, is accelerated from rest at $5\,ms^{-2}$ for 6 seconds. Determine the de Broglie wavelength for the electron at the end of the acceleration.

5 For the transitions shown on the diagram indicate for each whether photons are emitted or absorbed and calculate the frequency of each associated photon ($h = 6.63 \times 10^{-34}\,Js$).

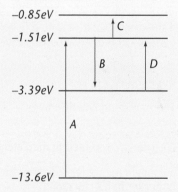

$-0.85eV$
$-1.51eV$
C
B D
$-3.39eV$
A
$-13.6eV$

Answers

1 frequency/electrons/kinetic/photon (particle ok) **2** a) KE = maximum kinetic energy of ejected electrons/h = Planck's constant/f = photon frequency/hf = photon energy/φ = work function of surface. b) $1.16 \times 10^{-18}\,J$ or 7.25 eV c) $1.9 \times 10^{15}\,Hz$ **3** wave: interference/particle: photoelectric effect **4** $2.4 \times 10^{-5}\,m$ **5** A: photon absorbed: $2.9 \times 10^{15}\,Hz$/B: photon emitted: $4.5 \times 10^{14}\,Hz$/C: photon absorbed: $1.6 \times 10^{14}\,Hz$/D: photon absorbed: $4.5 \times 10^{14}\,Hz$

If you got them all right, skip to page 77

Quantum Physics

Learn the key facts

1 There is plenty of evidence to suggest that electromagnetic radiation (e.g. light) is best described by a wave model. However, there is evidence to suggest that sometimes a particle model is appropriate.

The photoelectric effect:

When a metal surface is irradiated by electromagnetic radiation (usually ultraviolet) electrons are emitted from the surface of the metal (photoelectrons). The following observations are made.

em radiation

electron

A threshold frequency exists, i.e. there is a minimum frequency of radiation required before emission of electrons will occur. The value is dependent on the type of metal irradiated.

If the frequency of radiation is below the threshold then emission will not occur even when the intensity of the radiation is greatly increased.

The emitted electrons have kinetic energies that range from zero to a maximum value. Increasing the frequency of the radiation increases the maximum kinetic energy.

The maximum kinetic energy of the photoelectrons is independent of the intensity of the incoming electromagnetic radiation.

Electrons are emitted immediately the radiation arrives at the surface.

For single frequency radiation, the number of electrons emitted per second is proportional to the intensity of the radiation.

• A wave model alone fails to explain these observations. Increasing the intensity of the incoming electromagnetic radiation should increase the energy available to the electrons. The electrons would be expected to gain a greater maximum kinetic energy, as in a wave the energy would be distributed uniformly.

A wave model offers no reason for the energy of the emitted electrons being dependent upon the frequency of the incoming radiation.

A wave model does not account for the threshold frequency. With very low intensities of light, it should take a measurable time for an electron to accumulate sufficient energy to be liberated. This is inconsistent with the instant emission observed.

2 Electromagnetic radiation can be considered to be comprised of discrete packets (quanta) of energy known as photons. The amount of energy carried by a photon is found to be proportional to its frequency.

$$E \, \alpha \, f \quad \text{or} \quad E = hf$$

$h = $ Planck's constant $(6.63 \times 10^{-34}\,Js)$

Einstein suggested:

Single electrons are freed from metals by single photons; that is, there is no accumulation of energy from several photons.

There is an amount of energy required to free an electron from the attractive forces exerted by its surroundings. This is called the work function (f).

He proposed the equation:

$$hf = KE_{max} - \phi$$

$hf = $ photon energy/J $\phi = $ work function of surface/J
$KE_{max} = $ maximum electron kinetic energy

The minimum photon energy required to liberate an electron would be that equal to the work function of the surface. The electron would then be emitted with zero kinetic energy, i.e.

$$hf = \phi \quad \text{leading to} \quad f_0 = \frac{\phi}{h}$$

f_0 is the threshold frequency

3 Does electromagnetic radiation (light) exist as a wave or a stream of particles? To completely describe the properties of electromagnetic radiation a wave and a particle model are required; it depends on the observations made. This is called wave particle duality.

Wave model: reflection, refraction, diffraction, interference, polarisation.

Particle model: photoelectric effect.

De Broglie suggested that as waves exhibit duality so does matter. This means that waves can behave like particles and that particles can behave like waves.

Interference effects have been observed for beams of electrons.

Electrons and other small particles, such as neutrons, protons and hydrogen atoms, have been diffracted through crystal lattices.

4 De Broglie proposed that the wavelength and hence frequency of a particle is related to its momentum, where:

$$\lambda = \frac{h}{mv}$$

λ = wavelength/m h = Planck's constant/Js
m= particle mass/kg v = speed/ms^{-1}

The degree of wave behaviour demonstrated by macroscopic bodies is negligible. Due to their large mass, they would have a very small wavelength.

5 Electrons in atoms have potential energy due to the electrostatic attraction between themselves and the positive nucleus. This energy is quantised, i.e. there are certain allowed energy values that the electron can have. These values of energy are called energy levels.

The lowest energy level is called the ground state and is given energy level number 1. Often energy levels are illustrated as horizontal lines. The diagram shows the energy levels for a single-electron hydrogen atom.

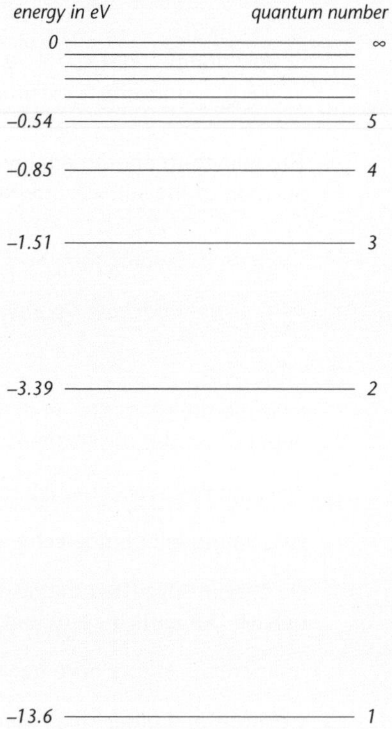

Energy levels are given negative values as a consequence of the selection of a reference point for zero energy. If an electron in level 1 were given 13.6 eV it would be liberated, i.e. there would be no potential energy as a result of interaction between the electron and the nucleus. The electron has been given energy but ended up with zero energy. It must have had a negative energy value initially.

Excitation occurs when an electron receives exactly the right amount of energy to promote it to a higher energy level. A photon would only achieve this excitation if its energy were equal to the energy difference between the higher and lower levels.

energy in eV   quantum number
0 ————————————————— ∞

−0.54 ————————————— 5

−0.85 ————————————— 4

−1.51 ————————————— 3

−3.39 ————————————— 2

−13.6 ————————————— 1

E.g. for the transitions shown in the diagram:

Transition A: $hf = E_2 - E_1$
Transition B: $hf = E_3 - E_1$
Transition C: $hf = E_4 - E_2$

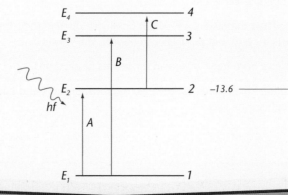

The atom will not remain in its excited state; after a short, unpredictable, time the electron will return to its ground state. E.g. an electron in level 4 may return directly to level 1 or it may return via other energy levels, 4 to 3 then 3 to 1 for instance. Each time the electron moves from a higher to lower energy level a photon is emitted. The photon has an energy equal to the energy difference between the two levels.

Ionisation occurs when an electron receives sufficient energy to liberate it from the confines of the atom. In the case of the hydrogen atom, the minimum energy required for the atom in its ground state to be ionised is 13.6 eV. Any energy received in excess of this will be carried by the electron as kinetic energy.

An atom can only emit certain photon frequencies. The frequencies are determined by the differences between energy levels; since only certain energy levels are allowed, this restricts the resulting photon producing transitions. The frequencies emitted are characteristic of the element, and all atoms of that element will emit the same frequencies.

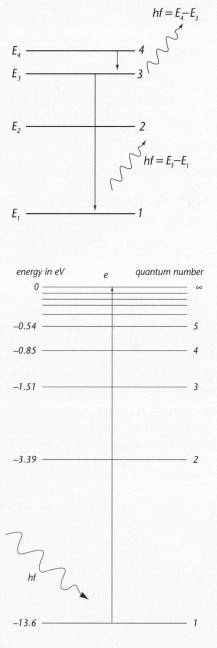

The emitted light can be diffracted using a diffraction grating. The angle of diffraction can be determined using a spectrometer, as shown in the diagram. The light is diffracted through certain angles into a series of discrete bands, called emission line spectra, which are viewed through the telescope. The angle of diffraction is related to the wavelength of the photons, where:

$$n\lambda = d\sin\theta$$

n = order of diffraction (1, 2, 3 …)   λ = wavelength
d = separation of diffraction grating lines   θ = angle of diffraction

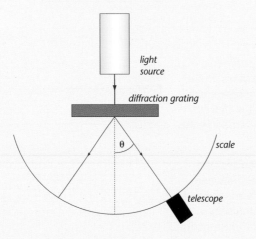

Determination, in this manner, of the wavelengths of light emitted from atoms provides very strong evidence for the existence of atom energy levels.

The spectrum for atomic hydrogen consists of three very distinct lines. These lines arise from the series of transitions shown in the diagram.

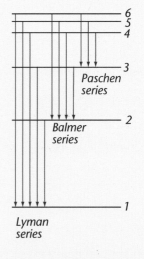

Have you improved?

1 Monochromatic electromagnetic radiation falls upon a metal surface, and electrons are ejected. What would be the effect on the maximum electron kinetic energy and the number of electrons ejected per second of the following?
  a) Changing the frequency of the incident radiation.
  b) Changing the intensity of the incident radiation.
  c) Changing the metal for one of a greater work function.

2 De Broglie suggested that all matter can exhibit wave-like properties. By estimating the wavelength of your body explain why you do not suffer the effects of diffraction when you walk through a doorway.

3 An atom has the energy levels shown in the diagram. It has two electrons, one of which is sitting in the $2\,eV$ level and the other at $8\,eV$.

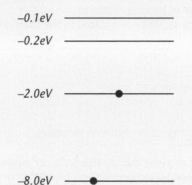

Describe the events that could possibly take place following the absorption of a
  a) $4.6 \times 10^{14}\,Hz$ photon
  b) $1.45 \times 10^{15}\,Hz$ photon.

4 Light of frequency $4.5 \times 10^{15}\,Hz$ falls upon a metallic surface of work function $4.8 \times 10^{-19}\,J$ causing the emission of electrons with maximum kinetic energy $2.46 \times 10^{-18}\,J$.
  Using this information calculate a value for Planck's constant.

1

2

3

4

5

DAY

6

7

The Nucleus and Radioactivity

How much do you know?

1 In Rutherford's experiment, which established the _____ model of the atom, he fired - _____ particles at gold foil. Since _____ passed through _____, he concluded that the atom comprised mostly _____ _____.

2 $^{238}_{92}$ U represents the nuclide uranium238. The top number is called the _____ number: it represents the number of _____ and _____ in the nucleus. The bottom number is the _____ number, representing the number of _____. All nuclei are _____ charged and contain the vast majority of the atomic _____.

3 Calculate for $^{235}_{92}$ U: a) The mass defect.   b) The binding energy.

4 Determine by calculation which is the more stable: $^{14}_{6}$ C or $^{92}_{36}$ Kr

Masses:
U235 = 235.0439 u
neutron = 1.0087 u
proton = 1.00728 u
1 u = 1.661 × 10⁻²⁷ kg
c = 3 × 10⁸ ms−1

5 a) Name the three types of ionising radiation.
   b) Which type of ionising radiation is emitted in the following decay?
       $^{224}_{88}$ Ra → $^{220}_{86}$ Rn
   c) Determine the values of X and Y $^{X}_{5}$ B → $^{12}_{Y}$ C + β

6 List α, β, γ radiation in ascending order of:
   a) Ionising ability.   b) Speed.   c) Mass.

Masses:
C14 = 14.003 u
Kr92 = 91.9264 u
neutron = 1.0087 u
proton = 1.00728 u
1 u = 1.661 × 10⁻²⁷ kg
c = 3 × 10⁸ ms−1

7 Give two examples of sources of background radiation.

8 If a sample of radioactive material initially has 3.7×10^7 unstable nuclei, how many would remain after 28 hours if the decay constant is 1.5×10^{-4} s⁻¹?

9 a) Calculate the half-life of a radioactive sample with a decay constant of 2.3×10^{-7} s⁻¹.
   b) If a sample initially contains N_0 undecayed atoms, how many would remain following four half-lives?

Answers

If you got them all right, skip to page 83

DAY

6

Learn the key facts

1 Rutherford (assisted by Geiger and Marsden) carried out alpha (α) particle scattering experiments to identify the structure of the atom. Prior to Rutherford's experiment, the 'plum pudding' model of the atom was generally accepted. This consisted of the positive charge in the atom being uniformly distributed throughout a sphere with a radius $\cong 10^{-10}$ m, and the negative electrons were thought to vibrate about fixed centres inside the sphere.

Rutherford demonstrated that this model was inaccurate and proposed the nuclear model of the atom.

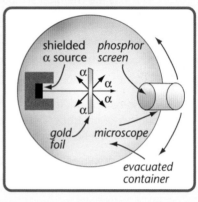

Basic method
A beam of alpha particles was fired at a thin gold foil.
The angle of deflection of the alpha particles was observed.
To avoid other collisions the experiment was carried out in a vacuum.

Observations
- The vast majority of the α-particles went straight through.
- A few were deflected though small angles.
- Fewer were deflected through angles greater than 90°.
- Even fewer (very rare) rebounded at 180°.

Conclusions
- Little or no deflection occurs for most alpha particles as a large amount of the atom contains only empty space.
- The large deflection angles experienced by very few of the fairly heavy alpha particles suggests that the positive charge in the atom is concentrated in a small region – the nucleus – and most of the mass of the atom is concentrated in this region.

In this nuclear model the positive charge is confined to a sphere of radius 10^{-14} m, the nucleus, and the electrons orbit this nucleus, occupying a spherical volume of radius about 10^{-10} m.

	mass/kg	charge/C
electron	9.11×10^{-31}	-1.6×10^{-19}
proton	1.67×10^{-27}	$+1.6 \times 10^{-19}$
neutron	1.68×10^{-27}	0

Within the nucleus are protons and neutrons. Protons are positively charged and neutrons have no charge, so overall the nucleus has a positive charge of equal magnitude to the charge on the electron. Protons and neutrons are collectively called nucleons.

Overall atoms are neutral as they have the same number of protons as electrons so their charges cancel.

DAY

6

2 It is the number of protons in a nucleus that gives an atom its identity. The number of protons in a nucleus is called the proton number (also atomic number), and the number of protons + neutrons is called the nucleon number (also mass number).

It is usual to write elements into nuclear equations in the form:

$$_Z^A Y$$

A = nucleon number $(n + p)$
Z = proton number (p)
Y = is the chemical symbol for the element.

When nuclear reactions occur, the total nucleon number and proton number on both sides of the equation must be the same. E.g.

$$_{92}^{235}U + _0^1 n \rightarrow _{56}^{144}Ba + _Z^A X + 2_0^1 n$$

for this reaction $A = 90$, $Z = 36$

An isotope of an element will have the same number of protons as the element but a different number of neutrons. Therefore, their atomic numbers will be the same, but their nucleon numbers will be different.

3 Einstein proposed that if a body changes its mass it will suffer a corresponding energy change. The two are related by the equation:

$$E = mc^2$$

If the mass of a nucleus is compared to the combined mass of the protons and neutrons that it is made from, then the mass of the nucleus is always found to be less than the mass of the constituent protons and neutrons. The difference in mass is called the mass defect, i.e.

mass defect = mass of nucleons − mass of nucleus

When protons and neutrons are combined to form a nucleus, energy is released. This release of energy removes mass in accordance with Einstein's equation. This energy is known as the nuclear binding energy (BE), where:

$$\text{binding energy} = \text{mass defect} \times c^2$$

If this amount of energy is released on creating a nucleus from its constituent nucleons, then the same amount of energy has to be supplied to break up the nucleus into its constituent nucleons. Hence the name binding energy.

4 Dividing the BE of a nucleus by the nucleon number gives the BE per nucleon. This reveals the relative stability of the nucleus. The greater the BE per nucleon, the more stable the nucleus (less likely to change). A plot of BE/nucleon against nucleon number is shown.

Nuclei near the peak of the graph have the greatest BE per nucleon and are therefore the most stable. ^{56}Fe is one of the most stable.

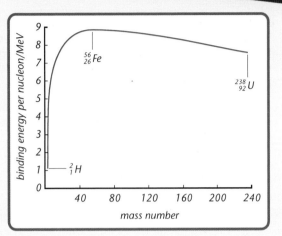

5 The nuclei of unstable elements disintegrate randomly and spontaneously into different nuclei of more stable elements.

The disintegration (decay) can result in the emission of ionising radiation which consists of one or more of: alpha particles (α), beta particles (β) and, less frequently, gamma waves (γ).

In general:

$$^A_ZX \rightarrow ^{A-4}_{Z-2}Y + ^4_2\alpha^{2+} \quad \text{alpha emission}$$
$$^A_ZX \rightarrow ^A_{Z+1}Y + ^0_{-1}\beta^- \quad \text{beta emission}$$

Emission of γ rays carry away excess energy when a daughter nucleus is created in a ground state. The proton and nucleon numbers are unchanged.

The disintegrating nucleus (X) is called the parent nucleus, the created nucleus (Y) is called the daughter nucleus.

6 The properties of ionising radiation can be summarised in the table below:

Type	Nature	Speed	Mass/kg	Charge	Deflecting fields	Ionising ability	Distance in air	Stopped by
a	He nucleus	0.06c	6.4×10^{-27}	+2e	Magnetic Electric	Very good	Approx 5cm	0.5mm paper
b	Fast electron	0.98c	9.1×10^{-31}	−1 e	Magnetic Electric	Good	Approx 500cm	0.5mm Al
g	em radiation	c	0	0	None	Very poor	several hundred m	Thick lead

A Geiger-Muller tube can be used with paper and aluminium foil to see what type of radiation a source emits. If an interposed piece of paper reduces the detected intensity significantly, then alpha particles are present. If the same happens when the aluminium foil is also added, then beta particles are present; and if the intensity is still higher than the background then gamma rays are also present; lead will stop these.

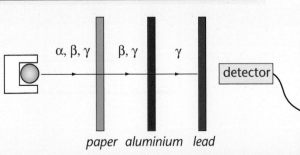

7 Background radiation is the radioactivity in the environment (i.e. everywhere!). It is around us all the time and must be taken into account when we carry out experiments with radioactive sources. Sources of background can be both natural (radon gas, rocks – especially granite – and cosmic rays) or artificial (mainly X-rays from the medical industry, contamination from nuclear power stations and weapons facilities).

8 The disintegration (decay) of nuclei is a random process but the more nuclei present in a sample, N, the greater the chance of detecting a disintegration. The number of disintegrations per second is known as the activity, A, of the sample.

$$A = -\lambda N$$

λ = decay constant (s^{-1})
– sign indicates that N decreases with time.

9 While the underlying nature of radioactive decay is random, we can make accurate predictions about the behaviour of a sample containing a large number of atoms, such as its half-life. Half-life is defined as *the time taken for half the unstable nuclei of a sample to decay.*

Half-life can be calculated using:

$$T_{1/2} = \frac{\ln 2}{\lambda}$$

The half-life of a sample containing one kind of atom is constant. A plot of N against t is an exponential decay curve.

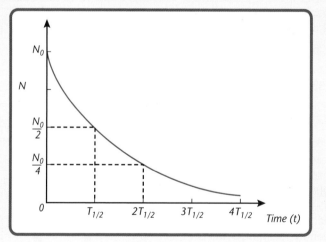

Have you improved?

1 Explain the meaning of the following nuclear expressions:
a) proton number
b) nucleon
c) nucleon number
d) mass defect
e) isotope.

2 How can the differences between the penetrating powers of different types of ionising radiation be used to identify their presence?

3 Sketch a graph with labelled axes of nucleon number against binding energy per nucleon. Indicate on your graph the positions of $^1_1$H, $^{56}_{26}$Fe, $^{238}_{92}$U.

4 The nuclei in a radioactive sample of 2.0×10^{40} polonium atoms decay according to the equation below:

$$^{218}_X\text{Po} \rightarrow ^{214}_{82}\text{Pb} + ^Y_2\text{Z}$$

a) Determine values of X and Y.
b) Identify the ionising radiation Z.

5 The count rate of a source emitting gamma radiation is found to decrease as the detector is moved away from the source. It is established that the measured count rate (C) is related to the separation between the source and detector (r) by:

$$C \propto \frac{1}{r^2}$$

If at $r = 2$ m the count rate is measured to be 5×10^{10} s^{-1} what would the count rate be 36 s later, 4 m away? ($T_{1/2} = 18$ s)

DAY

6

Physics of Particles

How much do you know?

1 Most of the mass of an atom is contained within its _____. This is not a _____ particle as it contains _____ and neutrons. Neither are these _____ as they can be split into smaller components called _____ which are _____.

2 a) Describe what occurs when an antiparticle collides with its corresponding particle.

b) A gamma photon of energy 1.02 MeV passes close to a nucleus resulting in pair production. Calculate the mass and determine the identity of each particle produced.

$c = 3.0 \times 10^8 ms^{-1}$

$m_e = 9.1 \times 10^{-31} kg$

3 What condition does a particle have to fulfil for it to be classified as a lepton?

4 a) What are the constituent particles of hadrons?

b) State two factors other than size that distinguish hadrons from leptons.

c) What are the three categories of hadrons? Give an example of each.

5 a) Describe the quark make-up of a meson.

b) Which types of quarks are in a neutron?

6 Which of the interactions listed to the right, if any, could occur? For those that do not occur state why not.

a) $\pi^- + p \rightarrow p + n$

b) $K^- + p \rightarrow \bar{K}^0 + n$

c) $\pi^0 + n \rightarrow K^- + \Sigma^-$

d) $p + n + e^- \rightarrow n + \bar{\nu}_e$

	Q	B	S
π^-	−1	0	0
p	+1	1	0
n	0	1	0
K^-	−1	0	−1
$\bar{K}^0$	0	0	−1
Σ^-	−1	1	−1
e^-	−1	0	0
$\bar{\nu}_e$	0	0	0
π^0	0	0	0

7 a) Name the four fundamental forces.

b) Two electrons separated by 2 m suffer a repulsive force. Which type of force, and via which exchange particle, is predominant?

c) How do you know that the answer is not the weak force?

8 If the Feynman diagram represents the reaction n + n → p + e⁻ via the exchange of a W⁺ particle, then what do [a], [b], [c], and [d] represent?

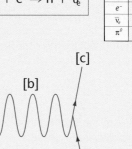

Answers

If you got them all right, skip to page 89

Physics of Particles

Learn the key facts

1 An atom consists of a nucleus orbited by one or more electrons. Is it reasonable to assume that the nucleus is a fundamental particle, i.e. one that cannot be split into smaller components? The nucleus is not the fundamental building block of nature; it contains protons and neutrons. It has also been established that both protons and neutrons can be split. They contain smaller particles, called quarks, which are thought to be fundamental.

2 It is useful at this point to consider the existence of the antiparticle. All fundamental particles and most others have a corresponding antiparticle. Particles and antiparticles have identical rest masses, but at least one other of their quantities will have opposite values, such as electrical charge or magnetic moment.

The table shows examples of particles and associated antiparticles.

particle		antiparticle	
name	symbol	name	symbol
electron	e^-	positron	e^+
proton	p	antiproton	$\bar{p}$
neutrino	ν	antineutrino	$\bar{\nu}$

Annihilation almost always occurs when a particle collides with its corresponding antiparticle, i.e. their combined mass is converted into energy. Other particles such as photons can be produced.

The mass is converted into energy according to Einstein's famous equation:

$$E = mc^2$$

E = energy; m = combined particle/antiparticle mass; c = speed of light

Pair production is the reverse of particle antiparticle annihilation. It is the spontaneous conversion of energy, i.e. a gamma photon, into mass. A particle and its corresponding antiparticle are formed. The combined particle and antiparticle mass is given by:

$$m = E/c^2$$

3 Leptons are a group of particles that are thought to be impossible to split into smaller components, i.e. they are fundamental. The family of leptons includes the electron, muon, tau particle and neutrino. There is, in fact, a neutrino associated with each of the leptons.
Each of these leptons has a corresponding antilepton.

name	symbol	charge	mass/kg
electron	e^-	–1	9.1×10^{-31}
muon	μ^-	–1	1.9×10^{-28}
tau	τ	–1	3.2×10^{-27}
electron neutrino	ν_e	0	1.3×10^{-35}
muon neutrino	ν_μ	0	4.8×10^{-31}
tau neutrino	ν_τ	0	6.2×10^{-39}

4 Hadrons are a category of particles that are not fundamental as they can each be split into smaller fundamental particles called quarks. All hadrons feel the strong force, i.e. the force that binds protons and neutrons in the nucleus. Leptons do not feel the strong force.

Baryon no.	Hadron subset	example
1	Baryon neutron (n)	proton (p)
0	Meson	kaon plus (K^+) pion plus (π^+)
–1	Antibaryon	anti proton ($\bar{p}$) anti neutron ($\bar{n}$)

In order to explain why certain interactions between hadrons do not occur, when by considering their properties they should, a property called the baryon number is introduced. Hadrons can be subdivided into three categories according to their baryon number.

Hundreds of hadrons have been observed, but the only one that is stable is the proton; all others will eventually decay into protons. E.g. neutron decay:

$$n \rightarrow p + e^- + \bar{v}$$

5 Quarks are thought to be the constituent fundamental particles of hadrons. This explains the properties of hadrons and their existence is supported by experiment: electron beams are used to examine the charge distribution within protons.

name	symbol	Baryon no.	charge	strangeness
up	u	+1/3	+2/3	0
down	d	+1/3	–1/3	0
strange	s	+1/3	–1/3	–1
charmed	c	–1/3	–2/3	0
bottom	b	–1/3	+1/3	0
top	t	–1/3	+1/3	1

There are six types (flavours) of quark. Some of their properties are illustrated in the table.

The top quark is predicted to exist but is yet to be discovered. All quarks have an associated antiquark.

Mesons consist of a quark and an antiquark (not necessarily the same type of quark).

E.g. pion = u$\bar{d}$ kaon = u$\bar{s}$

Baryons consist of three quarks, which can be of different types.

E.g. proton = uud   neutron = udd

During particle interactions, such as beta decay, a quark can change its flavour.

E.g. n $\rightarrow$ p + β + $\bar{v}_e$
    (udd) (uud)

A d quark in the neutron has changed (transmuted) to a u quark in the proton.

6 For all particle interactions that occur, certain properties (quantum numbers), e.g. charge, must be conserved. Conserved means that the total value before interaction must be the same as the total value after.

E.g. $A^{-2} + B^{+1} \rightarrow C^{+2} + D^{-1} + E^{-1} + F^{-1} + G^0$

total charge before $= -2 + 1 = -1$

total charge after $= +2 - 1 - 1 - 1 + 0 = -1$

therefore charge is conserved.

	Q	B	S
p	+1	1	0
π^+	+1	0	0
π^-	−1	0	0

Conservation of charge alone does not determine that an event will occur; all associated quantum numbers must be conserved. E.g.

$p + p \rightarrow p + p + \pi^+ + \pi^- + \pi^+$

charge (Q): $2 \rightarrow 2$ strangeness (S): $0 \rightarrow 0$ baryon number (B) $2 \rightarrow 2$

All quantities are conserved, so this could occur.

	Q	B	S
p	+1	1	0
K^-	−1	0	−1
π^-	−1	0	0
Σ^+	1	1	−1

$\pi^- + p \rightarrow K^- + \Sigma^+$

charge: $0 \rightarrow 0$ strangeness: $0 \rightarrow -2$

baryon number (B) $+1 \rightarrow 1$

Strangeness is not conserved so this could not occur.

7 Four fundamental forces act between particles. These are strong, weak, gravitational, and electromagnetic. All particles experience at least one of these forces.

force	experienced by
strong	hadrons (particles containing quarks)
electromagnetic	charged particles
weak	all particles (causes change in quark flavour)
gravitational	all bodies with mass

When particles exert a force upon each other, an exchange of an elementary particle (gauge bosons) occurs. E.g. when particles exert a gravitational force on each other due to their mass it is thought that the exchange of gravitons occurs.

force	exchange particle	range/m
weak	W^+, W^-, Z^0	10^{-18}
electromagnetic	photon	∞
strong	gluon	10^{-15}
gravitational	graviton	∞

8 Feynman diagrams pictorially represent interactions between particles. These diagrams illustrate the sequence of events and the involvement of exchange particles.

β^- decay $n \rightarrow p + e^- + \bar{\nu}_e$ β^+ decay $p \rightarrow n + e^+ + \nu_e$

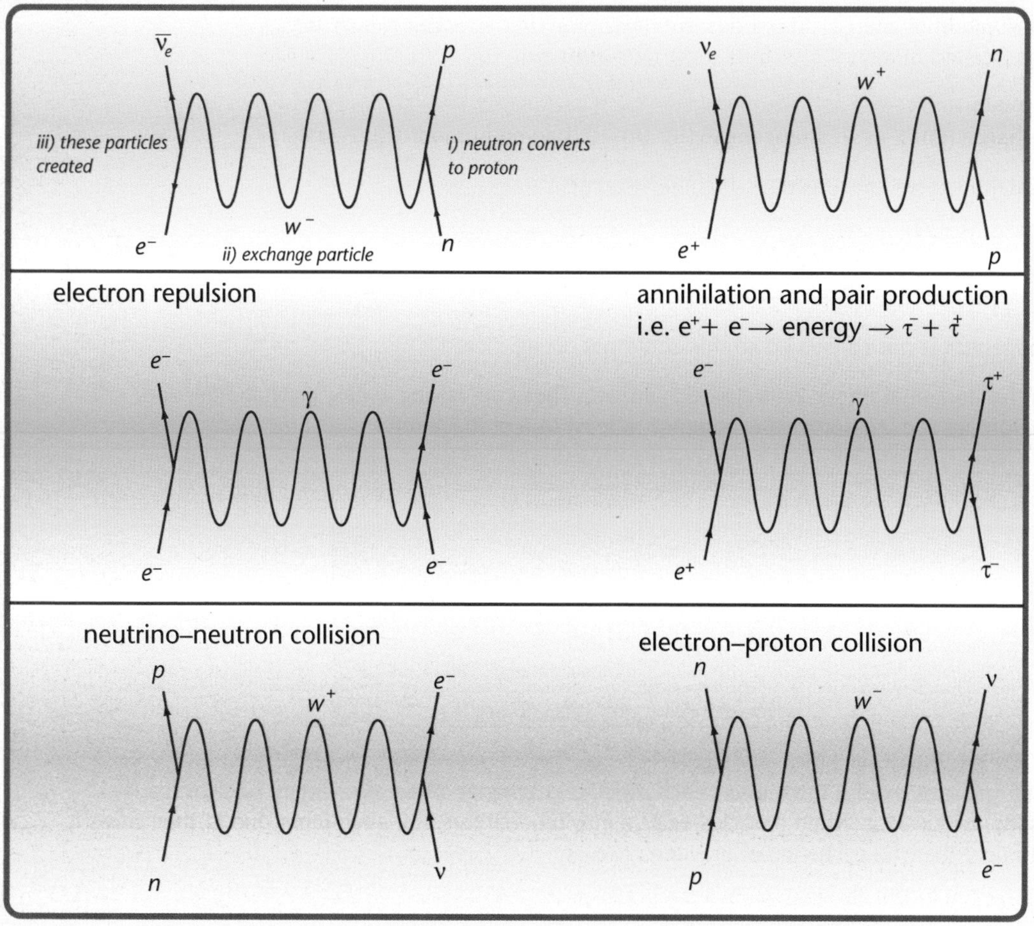

$\bar{\nu}_e$
p
iii) these particles created
i) neutron converts to proton
w^-
e^-
ii) exchange particle
n
ν_e
n
w^+
e^+
p
electron repulsion
e^-
e^-
γ
e^-
e^-
annihilation and pair production
i.e. $e^+ + e^- \rightarrow$ energy $\rightarrow \tau^- + \tau^+$
e^-
γ
τ^+
e^+
τ^-
neutrino–neutron collision
p
w^+
e^-
n
ν
electron–proton collision
n
w^-
ν
p
e^-

1
2
3
4
5
6
DAY
7

Have you improved?

1 The diagram below is a simple representation of a hydrogen atom (not to scale). For each of the boxes a) to f) identify all of the terms from the list that would serve as an appropriate label.

i) proton	v) antibaryon	ix) hadron	xiii) em force acts
ii) neutron	vi) pion	x) quark	xiv) nucleus
iii) baryon	vii) electron	xi) strong force acts	xv) fundamental particle
iv) meson	viii) lepton	xii) weak force acts	xvi) graviton

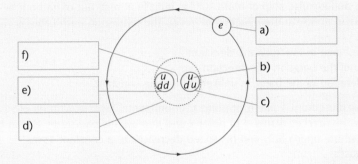

2 i) Write down the equations that correspond to the Feynman diagrams shown below.

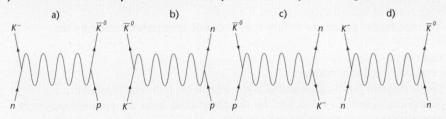

ii) By considering conservation of charge, baryon number and strangeness alone, determine which of the above reactions could occur.

	Q	B	S
p	1	1	0
n	0	1	0
$\bar{K}^0$	0	0	-1
K^-	-1	0	-1

3 A particle of mass M collides with its antiparticle, which is travelling in the opposite direction with the same speed. If pair production follows creating a stationary particle/antiparticle pair, each of mass $2M$, then determine, in terms of c (speed of light), the speed of the incoming particles. Comment upon your answer.

mass increase ΔM must come from the initial KE so $2(\frac{1}{2}mv^2) = \Delta Mc^2$

Exam Practice Questions

1 a) The list below gives some quantities and units. Underline those which are base units of the International (SI) System of units.

coulomb   Newton   mole   length   mass   temperature interval

b) Define momentum.

c) Use this definition to express momentum in terms of base units.

d) Explain the difference between scalar and vector quantities.

e) Is momentum a scalar or vector quantity?

2 Two tug boats are used to pull a cruise ship of mass 10^6 kg, initially at rest, out of harbour by two horizontal cables inclined at 30° to the direction of travel. The tension in each cable is 2500 N. Calculate:

a) the initial force on the barge in the 'forward' direction

b) the initial acceleration of the barge in this direction.

c) Explain why the acceleration of the barge is unlikely to remain constant.

3 A parachutist of mass 70 kg dives from a plane, at point X, in horizontal flight at altitude of 1000 m. After falling 100 m, he reaches $40\,\text{ms}^{-1}$ at point Y and deploys his parachute. Within the next 2.5 seconds, the parachute opens fully and his speed decreases to a constant $15\,\text{ms}^{-1}$.

a) Calculate the parachutist's:

i) loss in potential energy between X and Y

ii) kinetic energy at point Y, just before the parachute opens

iii) find the average resistive force between points X and Y.

b) It takes 2.5 seconds for the parachute to fully open. For this time period, calculate the parachutist's:

i) change in momentum

ii) the average resultant force experienced.

c) Draw a free-body diagram for the parachutist at the instant his feet hit the ground.

d) Draw a labelled graph of speed against time for the parachutist, from the moment he jumps from the plane until he comes to rest on the ground.

4 The centre of mass of a car of mass 1000 kg is 10m across a uniform bridge of mass 500 kg as shown.

(a) By taking moments about a suitable point, determine the magnitude and direction of the reaction force at support A.

(b) By resolving forces vertically, or otherwise, determine the magnitude of the reaction at support B.

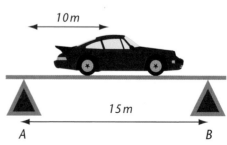

5 A 10V DC supply of negligible internal resistance is connected to a 1.2 kΩ ohmic resistor in series with a light-dependent resistor (LDR) which has a resistance of 800W in the light and 10kΩ in the dark.

a) i) What is the PD across the 1.2 kΩ resistor when the LDR is in the light?

ii) What current is taken from the battery at this time?

iii) What is the power dissipated in the resistor?

b) i) When the LDR is in the dark, what current is taken from the battery?

ii) How much charge passes through the LDR in 1 minute?

6 a) In a Young's slits experiment, light of wavelength 750nm was incident normally on a set of double slits, to produce a fringe pattern of separation 0.3mm on a screen at a perpendicular distance of 1m from the slits.

i) What general conditions for the light are required to observe interference fringes?

ii) Calculate the slit separation.

iii) What would the effect on the pattern be if (a) the slit separation were increased; (b) the screen moved closer to the slits; (c) the intensity of the light were increased?

b) i) Sketch the fundamental and first harmonic standing waves on a string of length 50cm fixed at each end.

ii) Calculate their frequencies (wave speed in the string is $120\,\text{ms}^{-1}$).

7 a) Write down Einstein's equation for the photoelectric effect, defining all terms shown.

b) Light of 500nm is incident on a metal surface of work function 1.8eV.

i) Write down the minimum KE of the liberated electrons.

ii) Calculate the maximum KE of the photoelectrons.

iii) What would be the effect on the KE of the emitted electrons if the intensity of the incident radiation were doubled?

iv) What would be the maximum KE of the photoelectrons if the wavelength of the incident EM radiation were doubled?

$c = 3.0 \times 10^8 \text{ms}^{-1}$

$h = 6.6 \times 10^{-34} \text{Js}$

DAY

7

Answers on page 96

Units and Quantities

1 Density/kgm^{-3}, Voltage/$kgm^2s^{-3}A^{-1}$, Pressure/$kgm^{-1}s^{-2}$, Impulse/$kgms^{-1}$, weight/$kgms^{-2}$

2 $F = k\rho v^2 r^2$; rearrangement gives:

$$k = \frac{F}{\rho v^2 r^2}$$

units: $F = kgms^{-2}$
$\rho = kgm^{-3}$
$v^2 = (ms^{-1})^2 = m^2s^{-2}$
$r^2 = m^2$

Therefore, units of k

$$= \frac{kgms^{-2}}{kgm^{-3}m^2s^{-2}m^2} = \frac{kgms^-}{kgms^-}$$

Units cancel so k has no units (dimensionless).

3 $A^2s^4m^{-3}kg^{-1}$

4
	$s = 2ut + at^2$			$s = ut + \frac{1}{2}a^2t$	
	↓ ↓ ↓			↓ ↓ ↓	
units:	m $(ms^{-1})s$ $(ms^{-2})s^2$			m $(ms^{-1})s$ $(ms^{-2})^2s$	
	↓ ↓ ↓			↓ ↓ ↓	
	m m m			m m m^2s^{-3}	

therefore it is homogenous   therefore it is not homogenous

	$R_T = \dfrac{R_1 + R_2}{R_1 R_2}$	$R_T = \dfrac{R_1 R_2}{R_1 + R_2}$
	↓ ↓	↓ ↓
units:	Ω $\Omega/(\Omega)^2$	Ω $(\Omega)^2/\Omega$
	↓ ↓	↓ ↓
	Ω Ω^{-1}	Ω Ω

therefore it is   therefore it is
not homogenous   homogenous

Equations that are homogenous with respect to units may be correct physical equations, however the numerical coefficients may still be incorrect. E.g. the first equation should be $s = ut + \frac{1}{2}at^2$

Vectors

1a) $231\,ms^{-1}$ b) $115\,ms^{-1}$

2a) $345\,N$ b) down

3a) $269\,N$ b) $112°$

4a) $4.2\,N$ b) $4.0\,N$ to the left

5 $5.6\,N$ at an angle of $36°$ up and to the left to the $10\,N$ force

Kinematics

1a) i) $11.7\,m$ ii) $1.5\,m$ in the direction A to B
iii) $6.9\,ms^{-1}$ iv) $0.88\,ms^{-1}$ in the direction A to B

b) $367\,ms^{-2}$

2a) To sketch a graph, start with what is known: the ball starts above the ground (taking 'upwards' as the positive direction) at point A and finally hits the ground (say this is where $s = 0$) at point C. How does the displacement–time graph vary between these two points?

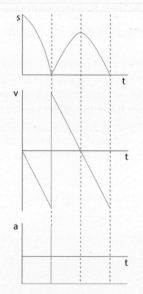

The gradient of the displacement–time graph is equal to the velocity: the ball starts from rest (gradient = 0) at A and falls 'downwards', which is negative by our choice of 'upwards' as positive (gradient is negative) – point B. The ball speeds up as it falls (gradient becomes larger and negative) – point C. Thus the velocity increases negatively from zero. After the ball hits the ground, it is moving in the opposite direction – upwards with a large positive velocity. So the gradient of the displacement–time graph is large and positive. The ball slows down (gradient becomes less positive) until it reaches its highest position. The graphs then repeat the same form as the ball once again falls. The acceleration is constant, due to gravity, apart from when the ball hits the ground, which exerts a large upwards force, resulting in a large upwards acceleration.

b) i) $-5\,ms^{-1}$ ii) 0.31 m iii) 0.5 s

3a) i) 80 cm ii) 4.0 m

b) i) $10\,ms^{-1}$ ii) $-4.0\,ms^{-1}$
iii) $10.8\,ms^{-1}$ at an angle of 22° below the horizontal.

4a) i) stationary ii) leftwards b) i) area between the curve and time axis ii) 0m c) i) gradient

ii) 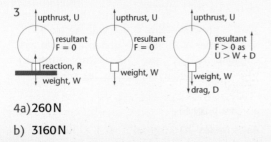 iii) change in velocity

Force and Momentum

1a) $6.7\,ms^{-1}$ in the initial direction of travel

b) $0\,ms^{-1}$

2a) 25000 Ns in the opposite direction to the exhaust gases

b) 5 kN in the opposite direction to the exhaust gases

3

4a) 260 N

b) 3160 N

5
a)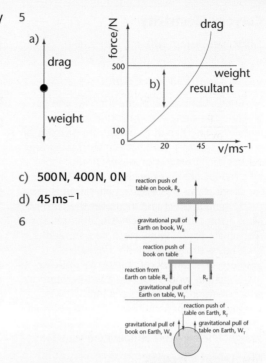

c) 500 N, 400 N, 0 N
d) $45\,ms^{-1}$

6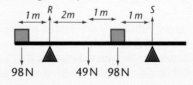

Equilibrium & Moments

1 613 N

2 1.5

3 a) 48.2° b) 13.5 N

4a) see diagram b) R = 171.5 N, S = 73.5 N

Work, Energy and Power

1a) 7.2 m b) 3.6 m c) 2.1 m

2 5 m

3a) 2 kW b) 8.3 kW c) 4 kW

4 a) 5000 N b) 90 kW

5 a) 11% b) 0.31 N

Current Electricity

1a) filament lamp b)i) a, f, j ii) c, d, h

2a)i) 15.4×10^5 C ii) 15.6^{-6} ms^{-1}
b)i) $1\,\Omega$ ii) $1.33 \times 10^{-3}\,\Omega$m

3   Some, but not all, of the thermal energy given to the sample is converted to atom vibrations and this does hinder the passage of current. However, some of the thermal energy causes the liberation of loosely bound electrons, which previously were not available for conduction. This electron liberation is the predominant effect and increases the current for a given pd, hence decreasing the measured resistance. (Note: while textbooks show this graph as a straight line with a negative gradient, this cannot, of course, be the case, as extrapolation would suggest a temperature at which resistance is zero, i.e. a high-temperature superconductor. The graph must therefore curve but it does approximate to a straight line for small temperature variations.)

Electrical Circuits

1a) 0.2 A b) 4 C

2a) 0.36 A b) 0.072 A c) 0.29 A d) 0.36 A

3   11.5 V

Basic Wave Properties

1a) The number of waves passing a point each second.

b) Transverse vibrations are limited to one plane only/ L waves do not oscillate perpendicular to the wave direction; they can only oscillate along the direction of travel of the wave/ any EM wave, e.g. light.

2   $v/2$

3a) speed of light, travel through a vacuum, transverse waves/frequency & wavelength

b) gamma ray/radio wave/X-ray/UV/microwave/visible/IR

4   Approx a) 0.7 m b) 30 m c) 500 nm

5a) 25° b) 65° c) 42°
d) as incident angle > critical angle, the ray is TIR and does not emerge.

6a) 81.28° b)

Further Waves

1

6 cm

2a) Path difference = 10 m = 5λ : constructive
b)   Path difference = 0 m = 0λ : constructive
c)   Path difference = 10 m = 5λ : constructive

3a) 8×10^{-8} m

b)   increase D or λ, or decrease the slit separation, s

c)   brighter fringes, and fewer fringes as there is less diffraction of the light through the slits, and so the interference region is smaller

d)   The central fringe is white, while all others are separated into the spectrum, with the shorter wavelengths (blue) closer to the centre than the longer wavelengths (red) for the first fringes. Fringes further away from the centre become 'blurred' as the different colours overlap.

4a) $\lambda_1 = 1$m/$\lambda_2 = 0.5$m/$\lambda_3 = 0.33$m/$f_1 = 250$ Hz/$f_2 = 500$ Hz/$f_3 = 750$ Hz

b) 2 m, 165 Hz; 2/3 m, 495 Hz; 2/5 m, 825 m

Heat

1a) 33.44×10^4 J b) 4.2×10^5 J
c) 1.25×10^6 J d) 0.55 kg

2a) P = pressure (Nm^{-2}) V = volume (m^3)
n = number of moles R = universal molar gas constant (8.31 JK^{-1}mol^{-1}) T = temperature (K)
b)i)   6017 moles ii) 193 kg iii) Releases 157 kg

3   See page 61

4a) 4 moles b) 7.5 gmol^{-1} c) 1×10^6 m^{2}s^{-2}

5   0.46 m^3

Solids

1

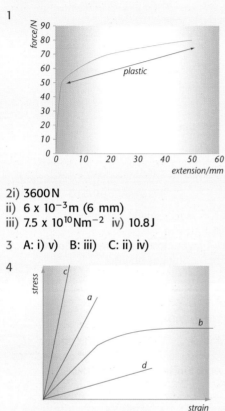

2i) 3600 N
ii) 6×10^{-3} m (6 mm)
iii) 7.5×10^{10} Nm^{-2} iv) 10.8 J

3   A: i) v)   B: iii)   C: ii) iv)

4

Quantum Physics

1a) Increasing frequency will increase the maximum kinetic energy.
Decreasing frequency will decrease the maximum kinetic energy to zero at the threshold frequency.
Changing the frequency will have no effect on the number of electrons emitted per second as long as the threshold is exceeded.

b) Changing intensity has no effect on the maximum kinetic energy. Increasing intensity increases the number of electrons ejected per second, proportionally.

c) A metal of greater work function would result in a lower maximum kinetic energy. The number of electrons emitted per second would be unchanged, providing the threshold remains exceeded.

2   $\lambda \approx 10^{-36}$ m. A doorway width of this order would be needed to diffract. You would not fit through.

3a) promote the -2.0 eV electron to -0.1 eV level

b) either, eject the -2.0 eV electron (ionisation) or promote the -8.0 eV electron to -2.0 eV

4   6.5×10^{-34} Js

The Nucleus and Radioactivity

1a) proton number: the number of protons in the nucleus b) nucleon: protons and neutrons are nucleons, i.e. particles from within the nucleus c) nucleon number: the number of protons and neutrons in the nucleus d) mass defect: the difference in mass between a nucleus and its constituent nucleons
e) isotope: an isotope of an element will have the same number of protons as the element but a different number of neutrons

2   Place a piece of paper between source and detector; if this reduces the detected intensity significantly, then alpha particles are present. If the same happens when the aluminium foil is also added, then beta particles are present; and if the intensity is still higher than the background then gamma rays are also present.

3   See diagram on page 81.

4a) $X = 84$, $Y = 4$
b) Alpha particle

5   3.1×10^9 s^{-1}

Exam Practice: Answers

Physics of Particles

1a) vii, viii, xv

b) x, xv

c) i, iii, ix

d) xiv

e) ii, iii, ix

f) xi

2i)a) $n + p \rightarrow K^- + \bar{K}^0$

b) $K^- + p \rightarrow \bar{K}^0 + n$

c) $p + \bar{K}^0 \rightarrow n + K^-$

d) $n + n \rightarrow K^- + \bar{K}^0$ ii) reaction b)

3 $v = c\sqrt{2}$. Impossible as v is greater than the speed of light.

Exam Practice

1a) mole

b) product of the mass and the velocity of a body

c) $kgms^{-1}$

d) scalar: described by magnitude alone

vector: has both magnitude and direction

e) vector

2a) 4330 N

b) $4.3 \times 10^{-3} ms^{-2}$

c) Resistance (drag) from water will increase with speed, decreasing acceleration. The tugs will also have a maximum speed.

3a)i) 7×10^4 J ii) 5.6×10^4 J iii) 140 N

b)i)   −1750 Ns ii) 700 N upwards

c)
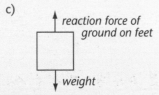

d)
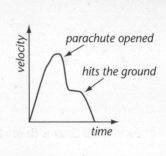

4a) 5833 N upwards

b) 9167 N

5a)i) 6.0 V ii) 5 mA iii) 30 mW

b)i)   0.9 mA ii) 0.05 C

6a)i) monochromatic, coherent
ii) 2.5×10^{-3} m
iii)a) fringes closer together b) closer together c) fringes brighter

b)i)

ii) 120 Hz, 240 Hz

7a) $hf = KE_{max} + \phi$; hf = photon energy, KE_{max} = maximum kinetic energy of photoelectrons, ϕ = work function of metal surface

b)i)   0 J
ii) 1.08×10^{-19} J (0.68 eV)
iii) no effect
iv) no electrons would be emitted as the frequency would be below the threshold.